WAYFINDING

ALSO BY M. R. O'CONNOR

Resurrection Science:
Conservation, De-Extinction and
the Precarious Future of Wild Things

WAYFINDING

—

The SCIENCE *and* MYSTERY
of HOW HUMANS
NAVIGATE *the* WORLD

M. R. O'CONNOR

ST. MARTIN'S PRESS
NEW YORK

www.stmartins.com

"The Experiment with a Rat," by Carl Rakosi, on page 173, from *The Collected Poems of Carl Rakosi* (Orono, ME: National Poetry Foundation, 1986). Permission granted by Daniel K. Nordby, Literary Executor for the Estate of Carl Rakosi (aka Callman Rawley).

Book design by Meryl Sussman Levavi

The Library of Congress Cataloging-in-Publication Data is available upon request.

ISBN 978-1-250-09696-8 (hardcover)
ISBN 978-1-250-20023-5 (ebook)

Our books may be purchased in bulk for promotional, educational, or business use. Please contact your local bookseller or the Macmillan Corporate and Premium Sales Department at 1-800-221-7945, extension 5442, or by email at MacmillanSpecialMarkets@macmillan.com.

First Edition: April 2019

1 3 5 7 9 10 8 6 4 2

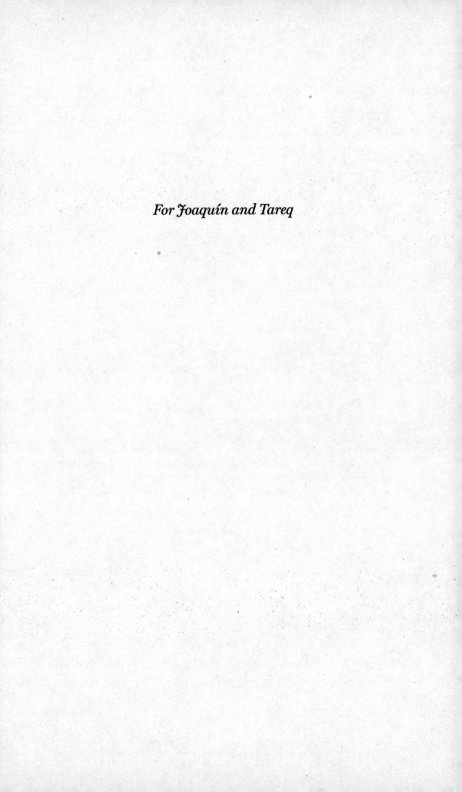

For Joaquín and Tareq

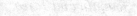

CONTENTS

There is no need to build a labyrinth
when the entire universe is one.

—JORGE LUIS BORGES

WAYFINDING

PROLOGUE
Wayfinding

In the high plains east of Denver we rented a car and drove due south on Interstate 25 straight as an arrow toward the cities of Colorado Springs and Pueblo. I watched the landscape rush by at seventy miles per hour, while overhead cumulus clouds cast their shadows on the open country. Just past the New Mexico border we turned southwest and began to hug the Santa Fe National Forest through Cimarron, then drove due west into Eagle Nest and Angel Fire. That night we slept in a motel in Taos and I woke up with an idea to visit a local hot spring on the banks of the Rio Grande. I put the name of the spring in my phone's navigation app and we drove out of town, following its directions, turning down a dirt road that led into a low-lying sagebrush *vega*. For the next while we turned onto one unmarked dirt road after another. I focused my attention on the phone's directions until I realized that the track had ended and we could go no farther. We got out of the car and met the waft of dust and crumpled sage, then walked fifty yards ahead to the edge of a cliff. I leaned forward and saw the Rio Grande surging a hundred feet below.

Somewhere nearby, I guessed, there must be a hot spring, and if only we had brought along some ropes and belay equipment or maybe a parachute we could have gotten to it with less risk to our lives. Though our predicament made me laugh, I started to wonder: what mathematical calculation based on an unknown, perhaps out-of-date map had come up with this murderous route? And why, I thought, had we naively trusted a disembodied algorithm and its satellite-radiated directions as it directed us toward a steep gorge? I had forgotten that my phone knew nothing of whether humans can fly, or the seasonal flow of the Rio Grande, that it had no actual experience because it had never been born, only programmed by someone who might never have set foot in New Mexico.

The novelist Audrey Niffenegger has written that there are different ways to react to being lost. Panic is one. Another is to surrender and "allow the fact that you've misplaced yourself to change the way you experience the world." We walked back to the car and sat on its warm hood. Severed from the umbilical cord of our GPS, we looked anew at the land. Before us lay a maze of brush stretching miles into the distance until it met the foot of the mountains, now cast in the purple shade of gathering thunderstorms. What was this place called? We didn't know and we had no map. We took in our unexpected perch and watched two distinct storm fronts to the north and south. The tangled balls of energy and lightning picked up speed, blowing toward us like tumbleweeds across the plain. The first drops of rain hit the dirt, and we raced our way out of the high desert labyrinth to a paved road that would deliver us to the sanctuary of better-mapped places.

For a long time I kept returning to that feeling of disorientation in New Mexico. I was struck by the power of a device to influence the way I moved through the world, how it subsumed my attention, mediated my perception, and lulled me into something like passivity. The way I viewed the technology in my hand changed; I felt suspicious. I was twenty-six when the first smartphone equipped with navigation technology was released, old enough that I'd spent my adolescence and the start of my adulthood relying on experience, habit, exploration, paper maps, signage, word of mouth, and trial and error to find my way around. I bought a smartphone in graduate school to get around the streets of New York City's outer boroughs as I hunted for stories and raced to cover breaking news as a newspaper reporter. Just a few decades before, the U.S. government had protected geolocation technology as a military secret. Now I had the power to know my latitude and longitude to within a hundred feet, velocity and direction to within a centimeter-per-second, and the time within a millionth of a second, giving me an imperious sense of mastery over my surroundings. Quickly—alarmingly quickly, in retrospect—my phone became *the* way I navigated, and I was not alone in my new dependence. In 2008, the year I got a smartphone, just 8 percent of American mobile phone owners used a navigation application to access maps and find their way; by 2014, 81 percent of owners were using them. In the period between 2010 and 2014, the number of GPS devices doubled from 500 million units to 1.1 billion. Some market projections expect that number to grow to 7 billion by 2022, mostly by expanding the use of GPS outside Europe and North America. Soon there could be a GPS device for nearly every person on earth.

Personal satellite navigation devices are the apotheosis of a dazzling era in human travel, an era of hypermobility. Most people have the ability to go where they want when they want, covering distances unimaginable to our ancestors at speeds that would have seemed proof of time travel just a hundred years ago. What was once an expedition is now a vacation. A voyage is now a jaunt. When the Venetian Marco Polo set off to the East in 1271, it took him four years to reach Xanadu and the empire of Kublai Khan in present-day China. He wouldn't see his homeland again for nearly two decades. In 1325 Ibn Battuta, one of the medieval ages' greatest explorers, set out for Mecca but ended up traveling as far west as Mali and as far east as China. It took him twenty-nine years. Technology has changed the very concept of a journey, a word that comes from the Latin for "diurnal," meaning a day's time. In Roman times the farthest one could travel in a journey was thirty or forty miles by horse. Since the start of the jet age in the 1950s, anybody who can pay the price of a ticket and possesses a passport can undertake what was considered a once-in-a-lifetime trip—what previously meant risking disaster, starvation, or death—in a day. There is joy in this freedom. Our reach is miraculous; our access unprecedented. But it's worth considering what, if anything, has been lost in the shrinking of space and time. The explorer Gertrude Emerson Sen, who founded the Society of Woman Geographers in 1925, questioned fifty years later whether her fellows' "travels today to the Arctic or the Antarctic or any other remote area, when you can fly there in a few hours, can be quite as fascinating as ours were in the olden days, when we travelled by slow freighters or camel, or on horseback or on foot."

Truly, the speed of change in how we relate to space and time has been scorching. We have turned roads into superhighways, flying into mass airline travel, locomotives into bullet trains; our cars may soon be self-driving. Marshall McLuhan believed that "after three thousand years of explosion, by means of fragmentary and mechanical technologies, the Western world is imploding. During the mechanical ages we had extended our bodies in space. Today, after more than a century of electronic technology, we have extended our central nervous system itself in a global embrace, abolishing both space and time as far as our planet is concerned."

We've had other eras of seismic change in how our species travels the earth. Our shift from mobile hunter-gatherer bands to sedentary communities and eventually states, what is known as the Neolithic Revolution some ten thousand years ago, has been described by Yale political scientist James Scott as a process of deskilling. At every step, he writes in *Against the Grain*, the skills necessary for survival "represent[ed] a substantial narrowing of focus and a simplification of tasks." If this seems too bleak a view of human civilization, he argues that at the very least our shift to sedentary livelihoods led to significant contractions: of our species' attention to practical knowledge of the natural world, of our diet, of ritual life, and of space itself. (The ancient Chinese, according to Scott, described nomadic people who registered with the state as having "entered the map.") During this period, our need to venture for hunting and resources likely shrank. Some trails and paths became roads connecting permanent settlements and greatly relieved the need to rely on memory and environmental landmarks to travel. As the scholar Alfredo Ardila writes, "For thousands of years, human survival depended on the correct interpretation of spatial signals, memory of places, calculation of distances, and so forth, and the human brain must have become adapted precisely to handle this kind of spatial information." Until recently, the vast majority of humans traveled without material maps.

Cheap and accurate GPS devices arrived in phones en masse just a decade ago, and already the era of paper maps and the challenge of orienting ourselves in space feels ancient. GPS seems indispensable, a psychic salve for getting lost or wasting time. Many of us embrace the device for even the shortest jaunts to ensure the fastest, most efficient route. In the *Boston Globe,* a journalist recounted a recent family road trip without GPS. Their adventures included using a telephone pole's shadow to tell west from east and identifying Polaris; it was a holiday exploring "the old ways." For those of us who remember the time before GPS, this lurch into a new normal feels abrupt, and the implications niggle at us. Weren't the old days . . . yesterday?

The pace of technological change sometimes makes it difficult to recognize the questions we should be asking. But in New Mexico I glimpsed a question: What happens when we outsource navigation to a gadget? Even the previous generation of navigation tools—the compass, chronometer, sextant, radio, radar—required us to give attention to our surroundings.

The pursuit of an answer led me into unexpected territory. What exactly is it that humans are doing when we navigate? How and why do we do it differently from birds, bees, and whales? How has the speed and convenience of technology changed how we move through the world and how we see our place in it? By drawing on research and insights from diverse fields of study—from movement ecology and psychology to paleoarchaeology, from linguistics and artificial intelligence to anthropology—I discovered a remarkable story about the origins of human navigation and how it influenced our evolution as a species. And I sought out individuals in three places—the Arctic, Australia, and the South Pacific—who practice what is sometimes called traditional or natural navigation, traveling great distances using environmental cues largely without the use of any maps, instruments, or gadgetry. For someone like me, who grew up surrounded by maps, this sort of navigation is a revelation, another way of looking at the world and thinking about space, time, memory, and travel.

We are a species of primate that shed our reliance on the biological hardware and genetic programming that tells animals where they are and need to go. Instead, we developed cognitive abilities built on perception and attention, giving us the freedom to go anywhere. For us, navigation is not pure intuition, but process. When we move through space, we perceive the environment and direct our attention to its characteristics, collecting information or, as some would describe it, building internal representations or maps of space that are "placed" in our memory. Out of the stream of information generated by our movement we create origins, sequences, paths, routes, and destinations that make up narratives with starting points, middles, and arrivals. It's this ability to organize and remember our journeys that gives us the ability to find our way back. More so, we mold the discoveries we make along the way into insights and knowledge that guide and orient us in our next explorations.

At the heart of successful human navigation is a capacity to record the past, attend to the present, and imagine the future—a goal or place that we would like to reach. In this way, navigation involves not only literal travel through space but also mental travel through time, what some call autonoetic consciousness. *Noetic* comes from the ancient Greek *noéō* meaning "I perceive" or "I understand," and the term *autonoetic* is now used to describe our ability to mentally represent ourselves as autonomous agents in time, giving us the capacity for self-reflectivity and self-knowing.

What cerebral anatomy makes this magic of consciousness possible?

Several regions of the brain are involved in spatial memory, including the parietal and frontal lobes. But neuroscientists have found that the principal place in the human brain responsible for navigation, orientation, and mapping is the hippocampus, a region of gray matter in our temporal lobe with a distinct ram horn–shaped curvature. If the firecracker-like firing of the hippocampus's different cells is stopped, humans lose the ability to find their way or recognize places they have been. People who have undergone trauma to or even removal of the hippocampus describe their waking experience as a kind of dream state in which their memories of locations and the events that take place at those locations disappear and every place and every experience is ever new. They lose their episodic memory, the ability to recall events of the past, and their capacity to formulate new memories essential for constructing a sense of self.

The hippocampus is critical for recording the what, where, and when of long-term memory in mammals. While there is debate over whether episodic memory is unique to humans or exists in other organisms, as far as we know, we are the only animals that can recall the events of our life and organize them into sequences to build identities. For our species alone, the hippocampus is the locus of autobiography, the narrative of the life we have lived till now. It is also the engine of our imagination: without it, people struggle to project themselves into the future, make predictions, or envision goals.

The hippocampus has sometimes been described as the human GPS, but this metaphor is reductive compared to what this remarkable, plastic part of our minds accomplishes. While a GPS identifies fixed positions or coordinates in space that never change, neuroscientists think what the hippocampus does is unique to us as individuals—it builds representations of places based on our point of view, experiences, memories, goals, and desires. It provides the infrastructure for our selfhood.

And the hippocampus is exuberant. The neuroscientist Matt Wilson has found that when rats fall asleep after running through a maze at his laboratory at MIT, the neurons involved in their internal spatial mapping system continue to burst with activity. By watching the pattern of their firing, Wilson can tell which part of the maze the rat is dreaming about. The animal's hippocampus is replaying the experience of moving through space. Wilson thinks sleep is likely a time when the hippocampus consolidates memories and seeks out rules and patterns of experience. "The idea is that during sleep you try to make sense of things you already learned," said Wilson. "You go into a vast database of experience and try to figure out new connections and then build a model to explain new experiences. Wisdom is the rules, based on experience, that allows us to make good decisions in novel situations in the future."

Why did nature so thoroughly intertwine spatial navigation and memory in humans? Which came first? The mysterious evolutionary story of the hippocampus hints at how we as a species differentiated from our nonhuman primate ancestors, and how that may have shaped our intelligence. The neuroscientist Eleanor Maguire has described navigation as a basic cross-species behavior because "the hippocampus is a phylogenetically old part of the brain, with an intrinsic circuitry that may have evolved to deal with navigation." But the late neuroscientist Howard Eichenbaum thought it was unlikely that navigation was the primary function of the hippocampus. After decades of designing maze studies on rats, he thought *memory* had always been the hippocampus's principal interest. "If navigation was the primary purpose, that would mean the chicken knew *how* to cross the road, but it didn't know why it was going there," he said. "I find it hard to believe that memory wasn't an adaptive feature that didn't come up earlier. For the chicken it's more like: there's good stuff across the road and now I need to get there." Eichenbaum described this neural circuit as a sort of magnificent grand organizer of the human brain, capable of mapping and sequencing multidimensional aspects of our experience in addition to space, from time itself to social relationships to music.

Research now shows that the volume of our individual hippocampus can be influenced by experience, and the size of this circuit has changed over time in our species as a whole. And while as far as we know the changes to our hippocampi created by our experiences are not passed down to our descendants, we do pass down genes that contribute to hippocampal volume; studies have shown that hippocampal volume is 60 percent heritable from parents to offspring. Nicole Barger at the University of California, San Diego, has found that the human hippocampus is 50 percent larger than would be predicted for an ape with the same size brain as us. Why is ours so much larger than that of other closely related apes? What selective pressures influenced the evolution of our hominin ancestors' hippocampi?

Maybe it has something to do with our ancient excursions. Humans are the only species to have inhabited every geographic niche; our distribution is remarkable compared to other life forms. What we don't know is what came first: did we travel far because we had big hippocampi, or did we get big hippocampi because we needed to travel far? What is for sure is that some fifty thousand years ago we fanned out from Africa. By twenty thousand years ago, our species had spread to Asia and Europe. By twelve thousand years ago, we had colonized the globe.

I like to try and imagine what this era of human travel was like—were our forebears constantly lost, inching hesitantly into unfamiliar places, ever fearful of the unknown, or were they like astronauts, each new generation pushing farther into the frontiers of topography and mind? Did we venture intentionally or drift through happenstance? Perhaps our cognitive powers gave us the tools to undertake these journeys, or maybe those long-range movements produced new strategies for navigation and then eventually the intellectual advances of culture and tradition that bound us in rich, emotional relationships to the places we called home. All of the aids we use today—roads, signage, maps, compasses, GPS—are nascent inventions in our species' history. For the majority of our species' existence, we traversed the earth using the landscape itself as a guide. And we seemed to have moved *a lot*.

Whether this movement took place over many generations or within individual lifespans is a matter of ongoing debate. It is difficult to model human movement patterns over a very long time with incomplete archaeological and paleoanthropological evidence. But it is interesting to consider whether there could be a part of us that is programmed to seek out and be curious. The various etymological roots of the word *seek* give a glimpse into this behavior's potential significance. In Sanskrit, the root *sag* means "to track down." In Latin, *sagire* means "to perceive quickly or keenly," and *sagus* "to predict" or "to be prophetic." These are all skills that would have been intrinsic to our success as a species engaged in foraging, hunting, socializing, and successfully navigating.

At the level of DNA, there is some evidence that a pattern of proteins in our genetic makeup expresses itself as an impulse to explore. In the late 1990s, Chuansheng Chen and several other researchers at the University of California, Irvine, started looking at dopamine receptor genes, especially an allele (a variant of a gene inherited from each parent, located on our chromosomes) called DRD4. Dopamine reception has been shown to influence exploratory behavior in animals as well as speed and vigor of locomotion. What the researchers wanted to know was if the presence of longer DRD4 alleles in individuals was caused by natural se-

lection through migration. If so, people from migratory populations would have a higher proportion of long alleles than those
from sedentary communities.

They looked at data from 2,320 individuals and found that the
length of this dopamine receptor gene correlated to the distance
they'd migrated from *Homo sapiens'* origins in Africa. These findings were controversial; dopamine correlates with characteristics
other than just exploratory behavior. But they provided some pro-
vocative ideas about the intertwined forces of genes and human
history. Jews who had migrated a longer distance eastward in the
direction of Rome and Germany had a higher proportion of long
alleles than those who went south to Ethiopia and Yemen. Bantu
individuals in South Africa, who had migrated from Cameroon,
showed a higher proportion too. The Sardinians, who live geo-
graphically closest to the origin of their language family, had
zero long alleles. Pacific Islanders, whose ancestors undertook
some of the greatest migration feats known to humanity, had
higher proportions of long alleles than any other group of Asians.
A later study by Luke Matthews and Paul Butler in 2011 focused
on the same DRD4 allele showed an even broader genetic profile
for what the researchers called novelty-seeking traits in humans,
including genetic quirks that might have predisposed our ancestors for exploratory behaviors—novelty-seeking and risk-taking.
(In contrast, biologists are discovering that chimpanzees' toler-
ance for novel stress is so low that transfer to new environments
and sanctuaries often leads to death.) Our rapid migration, Matthews and Butler hypothesize, selected for individuals who were
less vulnerable to novel stressors and whose risk-taking capacity
pushed them to explore.

Similar forces of selection seem to apply beyond our species;
nature sometimes chooses for the most fervent and driven seekers,
producing animals propelled by impulses and longings and the
biological hardware to undertake epic journeys. Consider monarch butterflies that fly south to central Mexico on stained-glass
wings for the winter and then return twenty-five hundred miles

north to feed and lay eggs on milkweed plants. In 2010 biologists discovered that female monarch butterflies who search farthest afield and most diligently for plants lay a higher amount of eggs. Tens of thousands of monarch butterfly life cycles have created organisms whose DNA seems infused with the compulsion to go farther. Monarch migrations cover so much distance that they eclipse a single lifespan. Butterflies that depart on the expedition die along the way and it's their great-grandchildren who complete the journey. Biologists call this a one-way migration: individuals travel in one direction but the population as a whole completes a circle. How do these organisms' grandchildren, who have never made the journey before, know where to go?

The mystery isn't limited to lepidoptera. Fifty different species of dragonflies fly along routes that are so distant, the insects die before the end of the journey and their descendants complete it. Aphids, maybe the world's least-appreciated migrants, inhabit a host plant in the spring and summer before producing young that fly to a different plant to mate and produce eggs. When these new aphids hatch, they fly back to the *original* plant, their grandparents' home, even though they have never been there before. Scientists strive to understand the assortment of mechanisms that gives these organisms such navigational certitude. Meanwhile, I envy their lack of existential wavering, the ability to always know where they belong and how to get there.

———

Where biology has failed humans in preventing us from becoming lost, we have substituted culture. We invented systems of knowledge for organizing environmental information to orient ourselves and cultural mechanisms to transmit this knowledge to the next generation. Often, difficult monotonous landscapes in flux—deserts, seas, ice—resulted in extremely intricate systems, the mastery of which could require years of inculcation and experience. In such formidable environments, survival depended on utilizing perception, observation, and memory. Sun, sky, stars,

wind, trees, tides, sea swells, mountains, valleys, snow, ice, anthills, sand, and animals are all navigational cues when interpreted
in context. As the aviator Harold Gatty believed, "With nature as
your guide, you need never be lost."

Not far into my research it dawned on me that despite having
traveled around the world, my experiences navigating it were mi-
serly compared to what is humanly possible. From one culture to
the next, we grow up absorbing different mental models, traditions, and practices. We undergo an education of attention, as psychologist James Gibson described. These cultural contingencies
of navigation intrigued me. In childhood we are exposed to languages, landscapes, technologies, and socioeconomic processes
that affect how we think and see. Some of us are born into completely oral cultures, while others start learning an alphabet as
toddlers. Some of us are taught how to read the earth or water to
find north and south, while others learn how to navigate mazelike city streets by taking sequences of left and right turns.

In recent decades, anthropologists and psycholinguists have
taken note of an astonishing range of human navigation systems
and begun chronicling them. Urban Europeans, Arctic hunters,
seafaring canoe sailors, desert nomads—each use unique practices
and skills to orient and know where they are. There is diversity
even at local scales: among neighboring regions, islands, and
communities. The Russian anthropologist Andrei Golovnev has
found that the Nenets, an indigenous reindeer-herding community in northwest Siberia, navigate very differently from their
neighbors, the Khanty. As Golovnev explains, "For Nenets, navigating is like watching oneself from the sky as a moving dot on
the map whereas the Khanty recognizes a tree and follows this direction, then he notices a hill and goes toward this point, remembering every detail of his hunting ground." These different
strategies infuse other practices: Nenets will fix a broken engine
by sitting in front of it and imagining the steps to fix it before
starting. A Khanty person starts unscrewing nuts right away,
because their hands remember every aspect of the engine. Two

different ways of navigating are arguably different ways of inter-
acting with the world.

What if even the experience of being lost is culturally contin-
gent? What if GPS is a gadget that addresses a specific set of cultural conditions: a severance of the individual from direct
experience and generational knowledge of place? To be sure, GPS
can and is used toward wildly diverse and oftentimes creative or
life-saving ends. The international Confluence Project, for instance, aims to photograph every single intersection of latitude
and longitude in the world, and members use GPS to locate them.
Syrian refugees depend on GPS to flee from war and travel across
the Mediterranean to Europe. Many use GPS to extend their reach
and explore places to which they might never otherwise go. Navigation devices make vast reserves of distributed knowledge available to us in an instant. But, crucially, they never require us to
possess information in our own memory in the way that successful navigators have been required to do till now.

Years after being led astray in New Mexico, I realized how
alien getting lost is to some and how unnecessary a tool like GPS
is to them. In northern Australia I met a Jawoyn elder in her eighties, Margaret Katherine, whose childhood was spent walking on
her family's traditional country near the Mann River. At one point
I asked what she did when she became lost in the bush. She
laughed. She took my notebook and illustrated how the termite
hills always pointed north-south, how the stars showed the way
at night, and how all of the rocks, trees, gorges, and escarpments
were created by her ancestors who traveled the world in the
Dreamtime. Their journeys and landmarks were recorded in songs
that she learned and memorized throughout her life. In this place,
which struck me as unmarked and bewildering wilderness, it
would be nearly impossible to become disoriented because
everywhere was home.

Later on, Ken MacRury, a historian of the Inuit dog and an
accomplished dogsledder, described similar levels of deep familiarity with place among the Inuit. In his decades of traveling with

hunters in the Canadian Arctic, he noted, "They wouldn't get lost. And the dogs never get lost, never. Fifteen or twenty years ago, the old Inuit couldn't believe when people started getting lost. They couldn't believe it was possible."

The anthropologist Thomas Widlok explained to me that people tend to generalize navigation from the Western perspective, which is largely about individuals trying to chart and map unknown territory. In his years traveling with the San people of the Kalahari, he rarely if ever witnessed them not knowing where they were. "You drive out on the weekend to Yellowstone Park and then have to find your way in this alien place that you consider wilderness. We Westerners find it very difficult not to transpose or project this perspective of, 'let's conquer the world,' as if this was a human universal," he said. "Trying to chart an area that you don't know yet is actually a very specific historical situation. It's a skill that is useful for imperialists wanting to create colonies. GPS is also a very useful tool for going into unknown spaces." The fascination with exploring unknown places is a "different mind-set from those in Australia, the San, or the Arctic," Widlok continued. "They do not aspire to colonize the world and occupy places they've never visited. They are mobile but they are mobile in a restricted sense, they stay within a more or less defined cosmos. They are not going into unchartered territory. They are doing something quite different."

I set out to talk with individuals who practice this "something quite different." Many of these unique practices have been lost to time or severed through cultural assimilation, oppression, and the extinction of languages. Modernity can engulf local ways of being, redefine borders or create new ones, and circumscribe movement or open up entirely different routes. The gas-powered engine, the speed of machines along fixed routes, cartography, GPS, and settlement have all changed how people navigate, whether in the American Midwest or the South Pacific. In some places I found individuals and organizations who consider the revival and practice of traditional navigation to be a matter of self-determination and cultural survival. By talking with some of them, I hoped to better understand the value and significance of these practices in the era of hypermobility, to perhaps even experience what the writer Robyn Davidson deems to be real travel: "to see the world, for even an instant, with another's eyes."

———

There is no single term that can encompass all the different processes and systems of human navigation. There are ongoing and contentious disagreements in anthropology, neuroscience, and psychology about the processes involved in what we do and how we do it, debates that I explore throughout this book. Yet I think there is one word that comes close: *wayfinding*. In the simplest terms, wayfinding is the use and organization of sensory information from the environment to guide us. The geographer Reginald Golledge defined it as "the ability to determine a route, learn it, and retrace or reverse it from memory through the acquisition of environmental knowledge." In the deepest sense, it is a concept that offers a new way of thinking about our connection to the world.

Four hundred years ago, the French philosopher René Descartes strove to explain human perception and started with the theory that our souls can only be in direct contact with our brains and not the universe outside our heads. Perception, according to Descartes's model, is a mechanistic process, and the outside is imagined in our minds because it is an image created by a physiological process. This is the basis of Cartesian dualism, the idea that consciousness is nonphysical and the mind and body are fundamentally separate. It was centuries before the scientific dogma that perception is the result of mental operations was challenged.

Born in 1904, the American psychologist James Gibson was fascinated by visual perception but frustrated by the assumption that there is a dualistic distinction between physical and mental environments. Through his studies of automobile drivers and airplane pilots, Gibson came to the conclusion that perception and behavior are a single biological phenomenon, and both humans and animals *directly* perceive their environment in an act of knowing or being in contact with it. We are not minds stuck in bodies but organisms that are part of our environment. Gibson called his theory *ecological psychology* and it led to a new understanding of navigation.

Gibson described the process of navigation as detecting the layout of the environment from a moving point of observation. When a person moves from one place to another, there is an optic flow of what he called *transitions*, a continuum of connected sequences in what we see that could be a turn in the road or the crest of a hill. These transitions connect vistas that open our view. Transitions and vistas are what provide us with the information we need for controlling locomotion and navigation. "We are told that vision depends on the eye, which is connected to the brain," he wrote in *The Ecological Approach to Visual Perception*. "I shall suggest that natural vision depends on the eyes in the head on a body supported by the ground, the brain being only the central organ of a complete visual system. When no constraints are put on the visual system, we look around, walk up to something interesting and move around it so as to see it from all sides, and go from one vista to another. That is natural vision." Later in life, Gibson rejected the idea of a cognitive map in the brain, instead adopting the term *wayfinding* to describe spatial navigation. There was no separation between mind and environment, between perceiving and knowing; wayfinding was a way that we directly perceive and involves the real-time coupling of perception and movement. He dedicated his book *The Senses Considered as Perceptual Systems* to "all persons who want to look for themselves."

Today, a handful of anthropologists and psychologists have adopted Gibson's ecological psychology model of navigation, describing wayfinding as a little-understood aspect of our everyday embodied existence in space. Because it is how we access the world and build consensus about reality, it is especially meaningful today when our attention is continually seduced downward to our devices and inward to our individualness. Wayfinding, I now think, is an activity capable of engaging with and attending to places and nourishing relationships and attachments to them.

At a time of social change and ecological disruption, the possibility of this reengagement with our surroundings seems incredibly important. There may also be more pragmatic concerns. Just as the field of neuroscience is revealing the complicated, beautiful influence of the hippocampus on human life, it is also revealing what might happen when we indiscriminately adopt a technology that allows us to dim activity in this part of our brains by using it to give us turn-by-turn directions. A growing body of research combining insights into spatial cognition, memory, and aging points at the significant neurological effects of not flexing the hippocampus: it can decrease in volume over time and adversely affect how we solve spatial problems. In a series of studies in 2010, a group of researchers at Montreal's McGill University, for instance, reported that exercising spatial memory and orientation in everyday life increases hippocampal gray matter, whereas underuse of its functions in older adults may contribute to cognitive impairment. Atrophy in the hippocampus is strongly associated with myriad problems, including Alzheimer's disease, PTSD, depression, and dementia. (One of the researchers, Véronique Bohbot, told the *Boston Globe* that she no longer uses satellite-navigation devices to tell her where to go.) Could GPS's turn-by-turn function have a subtle and potentially insidious impact on our well-being over the long term? While there has been no study testing such a direct relationship, the scientific literature so far indicates a possibility that a total reliance on GPS technology could over time put us at higher risk for neurodegenerative disease.

Meanwhile, scientists who study childhood development increasingly see the ability to explore, play independently, and self-locomote as essential aspects of cognitive maturation, potentially spurring the vigor of memory and theory of mind. Yet children's freedom of movement, from Japan to Australia to Europe to America, is increasingly circumscribed by risk avoidance and a lack of access to the outdoors. For the smartphone generation, getting anywhere without relying on GPS is becoming as dissonant with everyday life as writing by hand or reaching for an encyclopedia to find a fact.

Harry Heft, the environmental psychologist who studied with James Gibson, told me that his experience with his students today is jarring. "It's only a slight exaggeration to say that they couldn't get anywhere if they didn't have a GPS. What I find alarming about that is that it seems related to my concern about them not knowing history very well. History is important to me because it gives me a sense of where I am. I don't think they have a strong background in history, and I don't understand how they know their place in the world. GPS is like a smaller example of that. I worry that my students are disoriented. I think it's very existential: one needs to have a sense of where one is." The British author and self-styled "natural navigator" Tristan Gooley told me that the disappearance of navigational traditions and their replacement with modern technology signifies a cultural and philosophical impoverishment. "By using a GPS to find our way instead of clues available in the world itself, we devalue the experience of traveling anywhere," he said.

Tim Ingold, an anthropologist, sees the technology-drenched means of travel today, with its relentless goal of greater efficiency and convenience, as a further commodification of our lives. "We want to go faster and faster because we don't think anything is happening during that time," he said. "One could put it quite politically: travel today is a condition of advanced capitalism. Time spent on an airplane is opportunity passed because you could be doing other things, like making money." Ingold, who teaches at the University of Aberdeen in Scotland, has written extensively about the role of movement in human life, and he told me: "Life is a movement that carries on in *real* time, and it's something to value."

Perhaps the idea that each daily commute, ramble, exploration, expedition, migration, journey, and odyssey we undertake in the minutes, hours, and years of our lives is charged with significance is a romantic conceit drenched in nostalgia about imaginary olden days, a bygone era of nomadism, walkabout, or pilgrimage.

Or maybe wayfinding is an activity that confronts us with the marvelous fact of being in the world, requiring us to look up and take notice, to cognitively and emotionally interact with our surroundings whether we are in the wilderness or a city, even calling us to renew our species' love affair with freedom, exploration, and place.

PART ONE

———

ARCTIC

THE LAST ROADLESS PLACE

It took five years of planning and fifty-six days of sailing for British privateer Martin Frobisher to "discover" Baffin Island in the Arctic in 1576. Propelled by jet fuel in a Boeing 737, it took me twelve hours to skip over 23 degrees in latitude and arrive in the very same place 440 years later. My view from the plane window over the town of Iqaluit was of impenetrable white clouds that hung so low the wheels of the plane were nearly touching the tarmac before I caught sight of the airport. A fortress of yellow Lego blocks, it is the busiest airport in the Inuit territory of Nunavut in the eastern Canadian Arctic, which adds up to approximately a hundred thousand passengers each year, about half the number JFK receives in a single day. We disembarked down a metal flight of stairs to a freezing wind blowing wet snow into our faces. Inside, near the single baggage carousel, I watched as families waited for giant coolers of frozen food and provisions that they'd purchased in Ottawa to come off the plane. I had been warned in advance of Iqaluit's food prices: $20 orange juice and tomato sauce worth its weight in gold. I struggled to lift my own duffel bag crammed with dried fruit, jerky, and boxes of soup; swung it over my shoulder; and shook hands in the lobby with Rick Armstrong, a thirty-five-year resident of the Arctic and director of the Nunavut Research Institute. Armstrong had kindly offered to put me up in his spare bedroom. I threw my bags in the back of his pickup truck and we set off across town.

Sitting in the crux of a massive bay near the mouth of a river, Iqaluit was once a traditional starting point for inland caribou hunting, a place where people would have begun a fifty- or sixty-mile summer trek over rocky tundra, their dogs loaded with meat and supplies, to reach migratory herds. Today the caribou herds are fewer and Iqaluit is full of trucks and cars even though the longest route from one end of town to the other, Armstrong told me, only takes twenty minutes to drive. Most of the roads are dirt, and few have names. People describe where they live by the numbers on their houses, given in the order of construction, since the town was first permanently settled in 1942.

Frobisher had sailed to the Arctic in search of the fabled Northwest Passage, but he didn't have much aptitude for ship navigation and called himself a "poor disciple" of its arts. His training before departing consisted of a six-week crash course from the English alchemist and mathematician John Dee, who also served as the queen's astrologer. Dee envisioned an apocalyptic new world order of English Protestantism in which the queen was King Arthur incarnate and he was Merlin, her wizard advisor endowed with magical powers, overseeing a British Empire.

With Dee's help, Frobisher bought every newfangled piece of navigational technology available in the sixteenth-century marketplace: twenty compasses and mysterious objects with names like the "Hemispherium," "Holometrum Geometric," and "Annulus Astronomicus." Dee taught him to use a parabola compass that measured the magnetic variation from true north, and a wooden instrument called a Balistella that could measure the altitude of the sun or North Star to establish a ship's latitude, how far north or south it was. When it came to longitude, the east-west range of the ship, Frobisher had to rely on dead reckoning, calculating one's position by estimating the direction and distance traveled, as it would be another two centuries before an English clockmaker solved the puzzle of establishing longitude at sea. To assist in this calculation, Frobisher's manifest included eighteen "hower" glasses that could measure time: sailors threw a log attached to a line from the ship and used the glasses to measure the time it took for the ship to pass the log, giving them an estimate of speed that would then be used to estimate the distance the ship had traveled and its east-west position. Last, on Dee's advice, Frobisher bought Gerardus Mercator's map of the globe produced in 1569, the first to dissect space into rhumb lines, constant-bearing sailing routes projected onto the plane of the map.

I had come to the Arctic because the landscape has hardly changed in the past four hundred years, or in the thousand years before that. It is one of the last roadless places on earth. Just a few hundred yards outside of town there are no houses, lights, cars, railroads, signage, or cell towers, just ice, snow, rocks, and combinations of these elements in jutting and cascading variations. Most of the common navigation skills that will get you by anywhere else are nearly useless in this environment. GPS only lasts as long as a battery, and it can guilelessly lead you along treacherous routes, across faulty sea ice, or into bad weather. The magnetic field tries to pull compass needles downward. Even natural cues are fickle. The stars disappear in summer. In winter, the sun rises in the south and sets in the north. Polaris is a trustless companion for a traveler; above the Arctic Circle, you *are* north. Landmarks change appearance from season to season as snow gathers or ice melts.

And yet, for thousands of years the Inuit have thrived in the Arctic as intrepid travelers and hunters. What mysteries of navigation allowed them to accomplish this? "When adventure does not come to him, the Eskimo goes in search of it," wrote the twentieth-century anthropologist and writer Jean Malaurie in *The Last Kings of Thule.* Malaurie described the first encounters between Europeans like Frobisher and the Inuit as meetings between a "so-called advanced civilization" and an "anarcho-communist society." But it was also an encounter between two very

different ways of experiencing space—between those interested
in claiming ownership over it for the state and those seeking to
know it. The Inuit survived the extreme environment by becoming intimately familiar with its geography. They traveled on foot,
dogsled, and kayak, visiting hunting and camping places according to the season in the same fashion as their ancestors who migrated to the Arctic from the Bering Strait. Movement and the
knowledge it created was necessary for survival, a dramatic endeavor in a place of complex, extreme, and fluctuating conditions.

The arctic archaeologist Max Friesen has said that the early
inhabitants of the Arctic, who likely arrived around 3200 BCE,
probably had a life unlike any other ethnographic group till then
with "extremely high mobility levels and active exploration of previously unknown areas." By 2800 BCE, these Paleoeskimos had
reached the central Arctic, and within a few hundred more
years, northern Greenland. They traveled by foot and kayak but
covered vast areas to hunt sea mammals, musk ox, and caribou
with harpoons and bow and arrows. Around 1000 BCE, they were
joined by the Neoeskimos, also called the Thule, who searched
for bowhead whales and metal, built large skin boats, and used
teams of dogs to pull sleds. They not only shared the Paleoeskimos' capacity for moving across huge distances, they superseded
them: while the Paleoeskimos migrated across the Arctic over
several centuries, Friesen believes that some Thule people, direct
ancestors of the Inuit, may have crossed it in a single generation.

I have been told that one can still find Thule tent circles (rocks
used to anchor tents made of animal skins on the tundra), old
weirs in rivers used to catch fish, and stone cairns for pointing to
or caching meat and fish. Some campsites were used for so many
centuries that they had their own smell that hunters could detect
and follow like a trail to find their way. The thread connecting previous generations to the living is in the landscape. "Even burial
grounds and the things they left behind bring back memories,"
described Leo Ussak of the Kudlulik Peninsula. "A *qallunaat*

[white person] would feel the same way if he or she were to see an old cabin of an ancestor—that is why *qallunaat* have got their museums down south. This is how the Inuit feel when they go out hunting and see the things that our ancestors left behind."

Incredibly, instances of epic migrations by the direct ancestors of the Thule continued into the modern era. The last known migration, from Cumberland Sound on Baffin Island to Etah in Greenland, took place in 1863 and was documented by the Inuit-Danish explorer Knud Rasmussen. The journey covered islands, fjords, and open ocean and was led by a shaman named Qitdlarssuaq, who most likely had heard about other Inuit living across the sea from visiting whalers. Qitdlarssuaq became entranced by this knowledge, sent his soul traveling to look for these Inuit, and confirmed their existence. He then convinced thirty-eight men and women to go on a journey with him. They took ten sleds loaded with hunting tools, kayaks, tents, and skin clothes and set off. Each time they encountered an obstacle, Qitdlarssuaq sent his soul in the air to get a bird's-eye view of the landscape and find the best route forward. It took six years of constant traveling before they reached Etah and met the Inuit of northern Greenland, where they shared tools and intermarried. It was the remaining members of this expedition and their children that Rasmussen met in Greenland on his own epic travels fifty years later.

On my first evening in Iqaluit I bundled up and wandered the southern end of town, threading my way through packs of kids throwing snowballs and riding bikes. Most houses had a snow-mobile or three parked out front, as well as a *qamutiik*, an open sled traditionally pulled by dog teams, nowadays built with plywood and hard plastic runners. I made my way to a road at the edge of the bay and followed it round a bend to the Unikkaarvik Visitor Centre, where a festival celebrating Canadian films was underway. People helped themselves to juice boxes and plastic cups brimming with popcorn, and I sat on the floor in front of a taxidermied walrus to watch the movies, including one written and directed by a young woman, Nyla Innuksuk, from the Chesterfield Inlet, a community of around three hundred people on the western coast of Hudson Bay. Innuksuk's film is based on an old Inuit ghost story, retold in modern horror film fashion. A young man goes hunting on the land and makes a camp near an abandoned *iglu*, breaking a taboo. In retaliation, a murderous spirit kills his dog and then comes for him. The suspenseful, bloody climax sent shrieks of fear and delight through the audience.

When I left the visitor center it was nine thirty at night, and a vivid blue light, still luminous enough to see by, had been cast over Iqaluit. I walked back by way of an old graveyard at the edge of town and stopped to look at dozens of wooden crosses sticking out of the frozen ground. On the hillside to the west of me I could see the headlights of snowmobiles driven by teenagers racing up and down the steep face of hard snow. Beyond was the expanse of sea ice leading out to the bay where Frobisher had first made landfall, and across from me was the massive Meta Incognita Peninsula, Latin for "unknown limits," named by Queen Elizabeth and still emblazoned on both atlases and Google Maps. My thoughts turned to the finale of Innuksuk's film. "The land does not change," the narrator said. "You have beautiful things now but you should not depend on them. If you lost them tomorrow, could you live on the land? Could you hunt?"

———

Solomon Awa is a giant.

He crushes rocks with his bare hands to make diamonds. When he laughs, satellites are knocked out of orbit. His blood is used for flu vaccines. The legends of his brute strength have led to the nickname "Chuck Norris of the Arctic," and his accomplishments have become so legendary that Nunavut residents invented a hashtag just for him, #awafeats. They include the time Awa left an *inuksuk*, a traditional stone structure, on the moon to greet Neil Armstrong, and when he started commissioning a fat white man in the North Pole to deliver presents to the children of the world. The truth is, Awa's real-life feats are sometimes only slightly more believable. He can build an *iglu* single-handedly in under forty minutes. He was once heading to the floe edge—where the open ocean and winter ice meet—with his brother to hunt when they lost their snowmobile in a ten-foot-wide crack in the ice. Stranded in fifty-mile-per-hour winds, they made a sailboat of their *qamutiik* using a tent and some oars and rode the contraption across the ice until they were rescued.

So I was somewhat surprised when I encountered a rather small man with ruddy cheeks and a graying goatee framing a beaming smile. I shook Awa's weathered hand without injury and we got into the cab of his baby blue pickup truck. I expressed how lucky I felt locating him during the year's busiest travel and hunting seasons. Timing my visit to Nunavut had been challenging. I needed to be there when sea ice still covered the bay, otherwise I wouldn't be able to go out hunting. The year before, a long southern wind had kept the ice sheet in place for weeks longer than usual, but I had worried the opposite would happen this year: the ice might break up in early spring, and I would arrive too late and wouldn't be able to travel at all. Rapid climate change has made weather predictions in the Arctic almost impossible. Somehow I had arrived at just the right moment—the sea ice was thick, and because temperatures were still below freezing, it was hard and fast enough to snowmobile and dogsled on.

What I didn't realize is how these perfect conditions would make it extremely difficult to *find* hunters. A frenetic mood had gripped Iqaluit's residents. The sun barely set before it was yanked back into the sky like a yo-yo. The light was insistent, piercing. No one seemed to sleep. Any day the geese would arrive, and there were ptarmigan to shoot and fish to be caught. The seals were plentiful. The chance to be out on the land wouldn't come around again for months after the snow melted. With this in mind, I had called Awa, a keeper of traditional knowledge and a revered community leader, and timidly asked if he could spare half an hour to talk about wayfinding. He erupted in laughter. "Do you have thirty years?"

We drove to his favorite restaurant, fittingly called the Navigator Inn, and sat down at a Formica table. He ordered a cup of hot black tea and no food; he told me he had country food—the traditional and beloved fresh meat that could be seal, fish, whale, or walrus—at home to eat for dinner. Then he began talking. If I wanted to understand how the Inuit wayfind on the land, he said, I had to understand that whereas most people *like* to go places, the Inuit people *have* to go places. This necessity means that they grow up differently: from a young age they are taught through extensive, direct experience how to travel on the land and survive. "I was born in a sod house," he told me. "My father taught that when you get stuck, he asked a question. What are we going to do? In my mind I thought *he* was going to fix it, but instead he asked me. 'What are *you* going to do?' He knows he can fix it. *He* knows where he is. But he wants me to think so I can learn," he said.

Awa was one of eleven children raised in Mittimatalik (formerly known as Pond Inlet), a hamlet named after a hunter who died in a storm at the farthest tip of Baffin Island, some eighteen degrees of latitude from the north pole. The type of teaching his father gave him, direct observation and experience on the land, was especially important for learning navigation. Awa told me a story. When he was around the age of nine he was about ten miles out on the sea ice hunting seals with a group of boys. A dense fog rolled in and they had to leave right away before they lost all visibility. His father drove the snowmobile and the boys rode behind in the *qamutiik*. After a while, Awa's dad stopped. "Do you know how we got here?" he asked them. The boys admitted they hadn't been paying attention. "We didn't know where we were," recalled Awa. His dad continued to probe them. When they had traveled to the hunting spot, where was the sun? Where did they see shadows on the ground? Were the shadows darker on one side? "That was how he taught us. He asked a question," said

Awa. "And my father never wanted me to go to elementary school. He never really wanted that."

———

There are arguably few people left whose transition from a nomadic life to modernity happened as quickly and abruptly as the Inuit's. In a twenty-year period from the 1950s to the 1970s, nearly the entire population shifted from growing up on the land to living in sedentary communities administered by the Canadian government. A cash economy, gas engines, telephones, televisions, airplanes, hospitals, schools, and grocery stores arrived in the north. In a historical blink the Inuit adopted a new language, diet, and transportation. The speed of this cultural transition happened so fast that the Inuit dog, required for survival and travel for thousands of years, nearly went extinct.

The first interactions with *qallunaat* likely started in the tenth century with Norse seamen, followed by the European expeditions, and then whalers in the seventeenth century. Whalers and fishermen brought new materials like metal knives and needles, rifles, and cloth. Christian missionaries began arriving and "civilizing" the Inuit through conversion; Anglicans and Catholics competed with one another to gain the most followers. The missionaries imported a syllabic alphabet in order to teach people to read the Bible. Hunters and families were instructed not to travel on Sundays. Shamanic beliefs and age-old practices such as facial tattooing for women were banned. These cultural disruptions from contact with the *qallunaat*—not to mention the introduction of alcohol and European diseases—were devastating. But several intrinsic aspects of Inuit life stayed constant. Christian or not, people still needed to travel on the land and hunt in order to survive.

It wasn't until the Hudson Bay Company, the English-owned trading business, started establishing trading posts throughout the eastern Arctic in the nineteenth century that these practices began to change. The HBC sold furs for European markets, and many Inuit began trapping animals in order to trade the pelts for tobacco, ammunition, and food. HBC posts became hubs of economic activity that drew Inuit families into their orbit, initiating a shift from seminomadism to a sedentary life. The fur trade also had a pernicious effect on the Inuit's relationship to travel. As the anthropologist Sarah Bonesteel has written, trapping was time-consuming and diverted people from traditional subsistence activities like hunting. In turn, this shift increased people's dependence on the food the HBC stocked, which only increased the need to trap to make money to buy the food.

In the 1930s fur prices plummeted and caribou populations dwindled. With many Inuit facing starvation, the Canadian government passed legislation giving itself legal responsibility for their survival. When World War II ended, the Department of National Health and Welfare began implementing paternalistic social engineering programs designed to assimilate the Inuit into white society, ostensibly for humanitarian purposes. Children were legally required to attend residential schools and many were taken from their families. In school they were punished for speaking Inuktitut. The government established permanent settlements and gave the Inuit houses, encouraging them to become wage-earning citizens in mining and labor sectors. Healthcare infrastructure was built, and women were encouraged to give birth to their children in hospitals. Tuberculosis patients were relocated to sanatoria in the south; the Canadian government believed it could administer treatment to the Inuit more efficiently if they were stationary.

Solomon Awa's parents experienced these historical events directly. His mother, Agalakti, was born on the land and had an arranged marriage to his father, Awa, at the age of thirteen. In the book *Saqiyuq*, Agalakti tells the anthropologist Nancy Wachowich how she gave birth to their first daughter, Ooopah, at fifteen. They lived by traveling, hunting, and trading at outposts for the next thirty years. While Awa hunted by dogsled, boat, and foot, his wife sewed caribou-skin clothing for her husband

and children. When Anglican missionaries came to one of their camps, Agalakti and Awa were baptized and became Apphia and Mathias, taking Awa as a surname. In 1961 the Canadian government began forcing the Awas to leave their children at residential schools so they would learn English and be prepared for wage jobs. "They had to, that was the law of the teachers, that every student had to go to school. All my children were so young when they went to school. It seemed as if they were getting younger and younger," explained Apphia in an interview from the 1990s. "We left them there, but we missed them very, very much when they were gone. We missed them so much!"

Solomon was their eighth child, and when he turned seven years old, his parents asked the government if they could keep him. They offered two of their other children in return, just so they would have one son who could go hunting and camping with his father. They wanted Solomon to learn the Inuit way.

When a government boat came to pick up the children at their camp outside Pond Inlet, the officials took Solomon on board to treat a burn on his hand, which he had received from a pot of seal soup. But when Agalakti's husband went to pick up Solomon, he instead found Solomon sitting in the classroom; they had sent him to school. "My husband asked the teacher if he could take his son out of the classroom, and the teacher said no, so they started arguing. They got into a big argument, and then my husband just took Solomon by his hand and walked him out the door. He was very, very angry. He didn't even stop to get Solomon's parka. My husband gave Solomon his snow parka on the way back to the camp." Solomon was the only one of his eleven siblings to grow up on the land.

———

In the Arctic the sun is a fickle navigational aid. Above the Arctic Circle, it disappears in the winter below the horizon and then never sets in summer. There is a short period in March when it rises in the east and sets in the west. In order to deduce directional information from the sun, one has to know its complicated, changing journey through the sky at each time of year, and though many hunters do, it is never relied on as a primary navigation tool. Instead, as Awa explained to me, they use an assortment of aids, often in coordination with each other, to orient themselves in any condition and any landscape of the Arctic. One of these tools is *sastrugi*, snow that is formed into wavelike drifts by the wind. Imagine you are around Sanirajaq (also called Hall Beach), Awa said. It's completely flat, there are no mountains or landmarks. But at the beginning of wintertime, new snow falls into soft mounds called *uluangnaq*, a word that means the shape of a cheek. Then the prevailing wind, called *Uangnaq*, blows from the west-northwest. *Uangnaq* erodes the snowdrifts, carving them into new shapes. "They are sticking out like this," Awa said and stuck out his tongue. The tip pointed downward. "It's called *uqaluraq*, a tongue-shaped snow. And it's pointing north. So you look down on the ground, and that's how you tell where do I want to go, west or east? You can cross them."

Awa picked up the packets of grape and orange jelly on the table and began to arrange them to demonstrate. Say you are going south, toward the ocean, he said, placing a grape jelly at the edge of the table. Or you want to go north, toward land, and he put another jelly packet down in the middle of the table. "When you leave the community, you watch the tongues, which way you are cutting across them. Straight up? Or across? Or cutting at ten o'clock? One o'clock? When you want to come back home, you do the opposite way. If you were cutting straight up when you are leaving, then you cut straight down when you return."

Different cultures have invented various ways to classify winds. In ancient Ireland, winds were given a color; a southwest wind, for instance, was *glas*, a blue-green that translates to "the color of sky in water." The Inuit classify different winds by their character and moods. *Uangnaq* is generally considered a female wind: it gusts and dies down, blows furiously and then disappears. It's these volatile whims that shape the snow into *uqalurait*, and because it is the coldest of winds, it hardens the snow too. Even if another type of wind blows for a while, the *uqalurait* keep their form, and if fresh snow falls to cover it up, you can find *uqalurait* underneath it, still intact. In the book *The Arctic Sky*, navigator George Kappianaq describes how the *Nigiq* wind is male. It blows steadily and evens out the ground. *Uangnaq* and *Nigiq* have a relationship. When she blows especially hard, he responds by smoothing things over. *Uqalurait* are extremely important for navigation in flat places like sea ice where there may be no landmarks. The shape of the *uqalurait* can be discerned even when a storm is blowing and the moon, stars, sun, and landmarks aren't visible. Around Mittimatalik, where Awa grew up, it's a more mountainous region, and *uqalurait* aren't as necessary. Around there, Awa said he learned to use the Big Dipper, called *Tukturjuit* in Inuktitut, as a celestial guide, following its position throughout the night in order to tell time and direction.

In addition to the wind, stars, and snow, Awa explains that there is often a logic to geography. For example, around Iqaluit, he told me, the ridges and valleys all run south into the sea. He can use them as a compass while also recalling their particular characteristics individually and therefore utilize them as landmarks. People from the south would find this hard to believe. The Arctic environment, even mountainous areas, looks bewilderingly homogeneous. The tallest trees are the size of bushes, and the vegetation is more or less always the same—patches of moss and lichen on rocks. One jumble of rocks is identical to the next. Telling the difference between one snow-covered valley and another seems impossible.

"You go up there and every spot of land looks the same," Awa concurred. "But they are not! If you watch very closely, this place has a big rock. This other place looks like the same, but it doesn't have the *same* big rock. You have to watch for the details." This skill, according to Awa, is the key to the mastery of navigation: the Inuit people have an ability to perceive the subtlest of variations in detail and commit a staggering amount of visual information to memory. "I myself and everybody else who has gone out to the same spot over time, we know exactly what it looks like. If you are living in a location, you are there every day, you are knowing it like the back of your hand. As the English say."

But then, he tells me, many hunters don't even need to see the details of a place to commit them to memory, they just need to *hear* them described. Awa told me a story about a friend who had to deliver food from Igloolik to Pond Inlet, a distance of 250 miles as the crow flies. He had never been to Pond Inlet before, but someone who had made the journey gave him a description of what he would see if he followed the correct route. "He was told if you go here, you're going to see that," said Awa. "Then you go along the side and follow the valley and the valley ends and you're going to see a couple of humps, you go over those humps. He was told the story, and he followed through to the sea ice at Pond Inlet."

I looked down at our table, strewn in jelly packets. "Okay, but do you think the Inuit and *qallunaat* remember things differently?" I asked him.

"We have a hundred megapixels of memory, not one," he said. "And it's because we were taught oral history. Our memory is way bigger. I have no scientific information about that. It's probably the way memory is stored. The words that were taught, the stories, we put them in our memory, the land and visuals, we put them in our memory. That is why they told us to look at the details. When you are traveling, you put that important spot in your memory."

Awa's dinner was waiting for him. As we got up and stepped outside into the sun-drenched, freezing air, I asked him, "Do your children like to go out on the land?" "Oh yeah," he replied, "they love it. Being out on the land lifts you up spiritually, emotionally, and physically. It gives you medication, or meditation, however you want to call it. I'll never stop."

MEMORYSCAPES

In 1818 the Scottish captain John Ross launched another polar expedition to the Arctic in search of the Northwest Passage. Like Frobisher, he carried with him the wealth of Western navigation tools. These had evolved out of concepts of space and time dating back to the Sumerians, who developed the lunar calendar and organized time into hours and minutes and seconds, giving way to the invention of clocks, telescopes, sextants, and cartographic tools and charts.

During the voyage, he met two Inuit hunters and showed them a basic chart of the area. While the men had never read a map before, they recognized every place between Igloolik and Repulse Bay several hundred miles away, and they showed him their own route to the ship on the map, a journey that had taken nine nights. Then they *expanded* Ross's chart, drawing the coastline to the west and north and filling it with capes, bays, rivers, lakes, and camps and describing their favored routes. While Ross was impressed, they seemed to think their skills were nothing special. They told Ross that if he *really* needed help finding his way, they knew others with even more knowledge. Ross's journal is a compendium of these sorts of exchanges, in which the Inuit he meets possess an inexhaustible supply of information about the land.

Similarly, William Parry, who also sought the Northwest Passage, praised the "astonishing precision" of Inuit-created maps in the 1820s and their ability to masterfully depict every twist and turn of the Arctic topography and coast. Without the help of a map made for him by a guide, Parry believed he never would have found a critical passage through the Fury and Hecla Strait. The American explorer Charles Francis Hall mapped the coastline of Frobisher Bay in the early 1860s by asking an Inuk man whose name was Koojesse to sit at the helm of the ship and draw a chart of the coast as they sailed it. Later, when the Danish explorer Gustav Holm spent two years traveling in eastern Greenland in the 1880s, he collected several maps of the coastline from an Inuit man by the name of Kumiti. They were unusual: carved pieces of wood depicting a jagged, complicated coastline. "All the places where there are old ruins of houses (which form excellent places for beaching the boat) are marked on the wood map," wrote Holm. "[T]he map likewise indicates where a kayak can be carried over between the bottom of two fjords, when the way round the maze between the fjords is blocked by the sea-ice." The maps are both tactile and visual guides, models in relief of the contours of the country that might have been meant to be felt by the fingers for directions.

Knud Rasmussen was equally surprised by the inland Inuit, who had never used paper or pencils before but were able to pick them up and draw accurate representations of the land and the best routes to get wherever he asked. He used one of these maps, drawn by a man named Pukerluk, to navigate several hundred miles of the Kazan River in central Canada. "The historical record and modern cartographic research both agree that most Inuit maps, extensively tested through a century of use by non-Inuit explorers and field scientists, were extraordinarily accurate renderings of the landscape as sensually perceived," writes the geographer Robert Rundstrom.

Similar exchanges of navigational knowledge were taking

place on the other side of the world. When the British captain James Cook sailed to the Polynesian island of Tahiti on the HMS *Endeavor* in 1769, he met a priest from the island of Radiate. His name was Tupaia, and he told Cook about the long sailing journeys his people took to faraway islands. Cook inquired into their navigational methods and recorded that "these people sail in those seas from island to island for some several hundred Leagues, the Sun serving them for a compass by day and the Moon and Stars by night." He asked Tupaia to draw him a chart, and once he had "perceived the meaning and use of charts, he gave directions for making one according to his account, and always pointed to the part of the heavens, where each isle was situated, mentioning at the same time that it was either larger or smaller than Taheitee, and likewise whether it was high or low, whether it was peopled or not, adding now and then some curious accounts relative to some of them."

Tupaia died from disease in 1770, but his chart became one of the most infamous in the history of navigation—mainly because no one could figure out its underlying logic. It included seventy-four islands spread over a region bigger than the continental United States—one-third of the South Pacific—but the spatial relationships between them didn't make sense to the Western eye; no matter which way the map was flipped and turned, no coordinate system could crack the system behind Tupaia's placement of the islands. For the next several hundred years, historians tried to parse the geographical relationships it depicted. As late as 1965, some historians believed that Tupaia probably didn't draw it because "a non-literate man was fundamentally incapable of projecting his geographical knowledge on a piece of flat paper."

So many first contact experiences seemed to befuddle European explorers. The cultures they encountered didn't have compasses, astrolabes, ballistellas, or hourglasses, but people could nonetheless find their way across challenging and unforgiving geography.

For a long time, a popular theory of navigation was that indigenous peoples found their way by unconscious intuitions because
they were closer to animals, whereas Europeans had lost these
powers in the course of evolution.

The roots of this idea go back to at least 1859, when Alexander
von Middendorff, a Russian naturalist, suggested that magnetism
might explain how birds migrated. Some scientists speculated that
this ability might also exist in children and "non-industrialists,"
who possessed an imbued sense of direction, an unconscious
instinct they relied on to wayfind. One British colonial in India
wrote in 1857, "In the flat country of Sind . . . where one finds
neither natural landmarks nor tracks, the best guides seem to
count entirely on a kind of instinct . . . which seems to be the result of an instinct similar to that of dogs, horses and other animals." When the English explorer Charles Heaphy went to New
Zealand in 1874, he described a Maori man, E Kuhu, as having
an "instinctive sense, beyond our comprehension, which enables
him to find his way through the forest when neither sun nor distant object is visible, amidst gullies, brakes, and drives in confused disorder, still onward he goes, following the same bearing,
or divorcing from it but only so much as is necessary for the avoidance of impediments, until at length he points out to you the
notch in some tree or the foot-print in the moss, which assures
you that he has fallen upon a track."

The year before Heaphy's expedition, the scientific journal
Nature called for submissions on the topic of this mysterious
ability, and none other than Charles Darwin wrote to the eminent publication. He cited the case of Ferdinand von Wrangel, a
German explorer, who had written about the Cossacks' ability to
stay oriented over great distances, "guided by a kind of unerring
instinct." As Darwin described, "[Von Wrangel], an experienced
surveyor, and using a compass, failed to do that which these

savages easily effected." Darwin guessed that "some part of the brain is specialised for the function of direction" and that though all men are able to dead reckon, the natives of Siberia do it to a "wonderful extent, though probably in an unconscious manner."

For Darwin, this unconscious dead reckoning was evidence that organisms "preserved useful variations of pre-existing instincts" and that the human brain had preserved the same abilities seen in animals (e.g., passenger pigeons) to find their way home over long distances. For those humans to whom this instinct was useful, it was strengthened and improved by habit. Although Darwin conceded that all humans had the ability to dead reckon, he sought to place the skills of "savages" into an evolutionary hierarchy of which white European culture could be the only apex. Therefore, navigational aptitude could only be explained as a result of their proximity to animals on the evolutionary tree, and, like animals, their skills and mastery of the environment were biologically endowed, unconscious, and instinctive. In the early 1900s, the term "sixth sense" was coined to describe how blind people could avoid obstacles, and it was also ascribed to groups who exhibited uncanny navigational skills.

But Darwin appears to have glossed over a key aspect of von Wrangel's account. While the German explorer did write that his Cossack driver seemed to be guided by instinct, he also described how it was years of practice that gave his companion, Sotnik Tatarinow, the ability to use his memory to maintain a plan for navigating "intricate labyrinths of ice" and making "incessant changes of direction" so that memory and observation compensated each other and he never lost the main direction. "While I was watching the different turns, compass in hand, trying to resume the true route, he had always a perfect knowledge of it empirically," wrote von Wrangel. Once they reached flatter plains of ice, he described how Tatarinow was able to use ice landmarks in the distance to maintain direction while using *sastrugi*, the same patterns created by dominant winds blowing the snow that Awa described, to stay oriented. "They know by experience at what angle they must cross the greater and the lesser waves of snow in order to arrive at their destination, and they never fail." When *sastrugi* weren't present, Tatarinow switched to using the sun or the stars. Thus the ability to navigate in a seemingly "blank" landscape was actually based on Tatarinow's memory of the tundra in concert with environmental cues for orientation. Traveling several hundred miles between settlements without the aid of maps or instruments was not just common practice, it was the only way people who lived there had traveled for generations. Their abilities were based on intimate knowledge gained through direct experience, tradition, and rational calculations. They didn't need a sixth sense.

The British zoologist Robin Baker points out in his book *Human Navigation and the Sixth Sense* that one reason the scientific belief in a sixth sense lasted for so long is that by the nineteenth century, people in western Europe had so many aids to navigation, such as maps, compasses, place-names, roads, and road signs, that they themselves had *forgotten* there were other strategies for navigating. This forgetting is remarkable because, as Baker wrote, these modern inventions had only been available to the masses for three or four generations *at most*. "Throughout most of human evolution, navigation without instruments had been the rule," wrote Baker. It took just a couple of centuries for people to forget that environmental cues can be just as accurate as maps and gadgets. This historical amnesia made non-European navigation practices seem that much more supernatural and mysterious. As the accomplished Australian navigator Harold Gatty wrote, "In our Western civilization, pathfinding and natural tracking are so little developed . . . that notwithstanding all differences of innate ability, the man who has learned the simplest secrets of reading nature's signs is bound to outstrip the inexperienced observer, however intelligent he may be, and can not only outstrip him, but frequently amaze him. Nature's signs can be read with a little practice by the average intelligent Western man just as clearly as if they were street signs."

———

Of the hundreds of accounts by outsiders exploring the Arctic and of the thousands of anthropological studies of the Inuit, few spend much effort on understanding their methods of navigation. It confounded me that even Rasmussen, whose travels and encyclopedic recordings of Inuit life took place over tens of thousands of miles and three decades, never wrote about wayfinding explicitly or in very much detail—though he himself must have employed some of the same skills during his travels.

The first significant account of Inuit navigation that I could find was published in 1969 and was written by a young geographer by the name of Richard Nelson, who lived in an Alaskan Eskimo village called Wainwright. Nelson was under contract with the U.S. Air Force to write a practical guide for servicemen about how to survive in the Arctic based on indigenous knowledge. He ended up producing a detailed book, *Hunters of the Northern Ice*, describing the prodigious skills he witnessed, from how the Eskimo hunted at floe edges to their observations of astronomical phenomena. The appendix includes ninety-five words to describe sea ice alone, and around twenty pages of the book are dedicated to navigation.

One of Nelson's earliest experiences in Wainwright was watching his companion patiently seduce a seal to come to the surface of the water by rhythmically scratching the ice with a knife, exploiting its curiosity, until he shot it with his rifle and pulled it onto the ice. "You see," he said to Nelson after, "Eskimo is a scientist." Over the coming year Nelson learned firsthand what he meant. Arctic hunters studied every aspect of the environment—how animals behaved and ecology worked, and all of the interlinking causative connections between the phenomena they observed. Nelson saw hunters use color to study sea ice, the feel of paw imprints in snow to hunt polar bear, or the constellation Ursa Major to tell time and orient. He was unequivocal about the fact that these skills could be explained by intelligence alone. "There is no mystical inherited 'germ' in the Eskimo's mind that allows him to sense the mood of an animal, to anticipate the fickle movements of the ocean ice, or to sense a change in the weather," he wrote. "What may seem unfathomable to us at first is often so only because we lack knowledge and experience."

Among the other skills Nelson witnessed were hunters' abilities to observe and memorize landmarks and analyze their spatial relationships to one another, much like Awa had described to me in our conversation at the Navigator Inn. Nelson described a forty-mile trip by dogsled during which his companion was able to keep a true line of travel through what, to Nelson, was featureless landscape. His friend found "shallow stream valleys infallibly when there seemed to be no indication whatever where they should be" and "knew the location of fox holes completely in 'the middle of nowhere,' in spite of the fact that they were not visible from over a hundred yards away."

When I read Nelson's account of this journey, I was struck by its similarity to another anthropologist working thirty years later at the other end of the Arctic. Claudio Aporta, an Argentinian academic with a fascination for maps, geography, and the north, was writing a doctoral dissertation on Inuit traveling in the community of Igloolik, an island in the Foxe Basin. The area around Igloolik is the definition of a polar desert; it gets about as much precipitation as the Sahara, but it's so cold that it is covered by snow and ice for much of the year. It is also extraordinarily flat; Igloolik's single hill measures less than two hundred feet high.

During his time there, Aporta became interested in how southerners visiting the Arctic always tend to describe it as featureless and see the landscape as absent of life, a sort of blank space. "It *is* to a degree featureless," Aporta told me, "but what is interesting is how people who live there have to develop ways to wayfind. People need environmental clues and to find concrete places within those environments." Over the spring of 2000 and 2001, Aporta went on dozens of trips with hunters to understand how they did this. One of these trips was with a hunter who had laid some fox traps over twelve square miles with his uncle and wanted to retrieve them. To Aporta, the land looked completely barren. But somehow the hunter found *all* of the fox traps hidden under the snow. Aporta was extremely impressed. Then the hunter mentioned that he had laid the traps with his uncle *twenty-five years earlier* and hadn't visited them since. Now Aporta was astonished. "How is it that precise locations can be identified, remembered, and communicated without the use of maps?" he wondered.

Aporta came to the conclusion that the Inuit didn't travel across the land randomly—they followed known routes. In most places around the world, routes have been indicated with roads and human-made landmarks, and these paths and locations are then laid out symbolically on maps. The geographer Reginald Golledge argued that some environments are more legible than others because they have spatial coherence or an availability of landmarks that make navigation easier. The Arctic environment has ephemeral qualities that prohibit permanence—ice melts, snow is blown by changing winds, rivers flow and then turn to frozen ground come winter. Landmarks are either few and far between or difficult to distinguish and impermeable, so legibility depends on sociocultural dimensions, the symbolic significance and meaning imparted to the landscape by the people who traverse it. What Aporta found is that certain routes for traveling across the land become favored over generations, and knowledge of these unmarked trails is passed down from person to person, family to family, community to community, not in the form of maps but as oral descriptions that are memorized.

"One of the main differences between routes used by Inuit in the Arctic and those used by most cultures in other geographies," wrote Aporta, "is that in the Arctic, routes remain and evolve in the social and individual memory of the people, become visible only in certain periods as tracks on the snow, and disappear from the landscape as the seasons progress." In the course of around two thousand miles that Aporta traveled on snowmobile around Igloolik, he mapped thirty-seven known trails, fifteen of which had been used by multiple generations of travelers. Over that same distance, he recorded over four hundred Inuktitut names for places. There was no point at which his companions didn't know where they were. Inuit routes don't go through a no-man's-land. Routes go "through named features, across patterned snow, and along familiar horizons, all of which constitute the territory in which a good traveller always knows where he/she is," he later wrote.

Aporta thinks that travel for the Inuit is rarely transactional or driven by survival, or even the need to get from A to B. Travel is a way of being. Babies were conceived and born along routes, people convened to share resources and news. Aporta now sees the Arctic landscape as an example of a "memoryscape," the mental images of the environment and places, remembered by individuals and shared among people. The term was invented by social anthropologist Mark Nuttall in reference to the Greenland Inuit in the early 1990s: "The relationships between a hunter or fisher and the environment are constituted in part by personal and collective memories that invest the landscape with personal, family and local significance, but also by ideas of rootedness and fixed attachment," Nuttall has written. Traveling along these routes was the way the Inuit engaged with the environment and maintained, nourished, and expanded their memoryscape.

———

Spatial orientation researchers generally break human navigation strategies into two sorts. The first is route knowledge, an ability to construct a sequence of points, landmarks, and perspectives that make up a path from one place to another. The traveler uses a string of memories of landmarks or viewpoints to recognize the correct sequence for getting from one place to another. Initially, the Inuit use of memoryscapes would seem to be a clear-cut example of route knowledge. The second strategy is called survey knowledge: the traveler organizes space into a stable, maplike framework, in which every point or landmark has a two-dimensional relationship to every other point. While route knowledge is the verbal description you might give when telling a friend how to get to the post office, survey knowledge is the "bird's-eye" map of the walk you might draw for that friend on a piece of paper.

Route knowledge relies on the traveler's point of view and relationship to objects around them, what's called *egocentric* perspective. The individual sees everything in relationship to themselves and their body's axes—in front, behind, up, down, left, and right. Survey knowledge depends on what is called an *allocentric* perspective, a point of view that is objective, maplike, and nonindexical in its representation of spatial locations of objects and landmarks.

Throughout the twentieth century, psychologists thought that the egocentric perspective was the most intuitive, simplistic, and primitive kind of spatial reasoning. Researchers like the Swiss psychologist Jean Piaget argued that young children possess an egocentric perspective first; only as they matured around age twelve did children develop the capacity for the objective vantage point of allocentric or Euclidean coordinate space—what he called the formal operational stage. But Piaget, who was sometimes called a "cartographer of the mind" by his peers, mainly studied small groups of European children, and his findings have since been criticized for being unrepresentative. Indeed, they may be a classic example of a long-standing problem in psychological literature: making broad claims about human psychology based on narrow samples taken from what are now called Western, Educated, Industrialized, Rich, and Democratic, or WEIRD, societies. Since Piaget's era, the simplistic progression from egocentric to allocentric knowledge in individuals has been disproved by psychologists like Charles Gallistel of Rutgers University, who has shown that individuals—even children—are often capable of utilizing *both* strategies: apprehending the environment from the visual flow of locomotion or using spatial cues such as those that come from surveying an environment from an elevated position.

As the psycholinguist Stephen Levinson has written, there is a similarly flawed tradition in linguistics of assuming that the language used to describe space is a reflection of universal egocentric spatial concepts. Immanuel Kant thought as much and argued that our intuitions about space are based on the planes of the body: "up" and "down," "left" and "right," "back" and "front." Presumably, some cultures then built on these innate, biologically endowed concepts through sociocultural inventions like charts, compasses, and clocks, the tools needed to organize space allocentrically.

Yet this version of cultural categorization has also been disproved. A huge variety of people and languages use an allocentric perspective and strategy, including those that produce no material navigation technologies. Clearly, the Inuit use memoryscapes but are more than able to accumulate and import survey knowledge of the land. So, just as individuals can use an amalgam of strategies to navigate, it's virtually impossible to assign universality to cultural spatial navigation strategies or language, let alone organize cultures into pure hierarchies and call them Eastern or Western, primitive or modern, scientific or preindustrial, egocentric or allocentric. "Thinking at the highest level, at Piaget's stage of formal operations," writes anthropologist Charles Frake, "is not, as many have claimed, the hallmark of the modern, literate, scientific mind, but is, rather the hallmark of the human mind when confronted with a task sufficiently necessary, sufficiently challenging, and sufficiently clear in outcome." Indeed, some so-called innate differences may have a lot to do with the topographies we inhabit. As the Colombian scholar Alfredo Ardila has posited, contemporary city life calls for the logical application of mathematical coordinates, whereas for much of human history people oriented themselves in nature by interpreting spatial signals and memory and calculating distances from environmental cues. Depending on where we are born, the languages we speak, and the topography we dwell in, it seems we are all capable of utilizing different cognitive strategies to greater or lesser flexibility and mastery in the task of navigating.

At the restaurant in Iqaluit, Awa told me he believes that the Inuit can navigate the Arctic because they have bigger memories than *qallunaat*, though he pointed out that he had no scientific proof of it. His insight strikes at the heart of the intriguing relationship between human navigation and memory. While neuroscience has only recently begun to reveal its physiological basis, this relationship fascinated even the ancient Greeks, who afforded great respect to individuals who could memorize vast amounts of information. In the Elder Pliny's *Natural History*, for instance, he recalls that Mithridates of Pontus knew twenty-two languages, and Cyrus knew all the men in his army by name. In the *Ad Herennium*, a Latin book from around 80 BCE, the unknown author (once thought to be Cicero) tells how Seneca could listen to two hundred students each recite a line of poetry and then recite all the lines perfectly—starting with the the last line and ending at the first. He could also supposedly repeat some two thousand names in perfect order after hearing them once. Another rhetoric teacher, Simplicius, could recite Virgil's *Aeneid* backward.

To aid memorization, the Greeks invented an art dedicated to it, the method of loci, a system that appears to have taken advantage of the human brain's proclivity for spatial memory to create an ingenious mnemonic device. In her 1966 book *The Art of Memory*, the English historian Frances Yates describes how it was Simonides of Ceos, the "honey-tongued" lyric poet, who invented this art of memory some twenty-five hundred years ago. Most of what we know about his system comes from just three texts in Latin, the *Ad Herennium*, Quintilian's *Institutio oratoria*, and Cicero's *De oratore*. It is Cicero who recounts how Simonides attended a massive banquet in Thessaly and, after reciting a lyric poem he had composed in honor of the host, was told by a messenger of the gods Castor and Pollux to go outside and meet them. But when he left, the roof of the hall collapsed and killed all the guests. The bodies were so badly injured that no one could identify them—except for Simonides. He remembered where each person had been sitting at the table.

What Simonides discovered through this experience was that by imprinting or stamping *loci*, or a place, in one's mind and placing a memory in that place, it could be easily recalled. He recommended that one build an architectural structure with rooms and hallways imagined in great detail and then put information, names, and words in those places. When the orator or person needs to recall a piece of information, he revisits the building and the places where he has stored his memories. When it comes to long pieces of lyric poetry or ballads, the author of the *Ad Herennium* instructs students to learn the verses by heart by repeating them, and then replace the words with images and associate those images with *loci*.

The method of loci was practiced by some of the great minds in Western history through the Renaissance, even after the invention of the printing press and ubiquity of the written word. Yates thought that these ancient memory systems were "at the great nerve centres of the European tradition" and that by the seventeenth century the art of memory was helping to midwife the age of scientific inquiry in Europe, helping scientists and naturalists to draw "particulars out of the mass of natural history, and ranging them in order. . . . Here the art of memory is being used for investigation of natural science, and its principles of order and arrangement are turning into something like classification."

Some ancient Greeks it seems were wary of how the transformation from oral to literate culture might affect their memory capacity; they viewed the written word with apprehension. In the *Phaedrus*, Socrates recounts the story of the Egyptian god Theuth, who enthusiastically reveals the invention of letters to King Thamus, promising him that it will improve memory. But Thamus thought that he was mistaken.

> For this invention will produce forgetfulness in the minds of those who learn to use it, because they will not practise their memory. Their trust in writing, produced by external characters which are not part of themselves, will discourage the use of their own memory within them. You have invented an elixir not of memory but of reminding; and you offer your pupils the appearance of wisdom, not true wisdom, for they will read many things without instruction and will therefore seem to know many things, when they are for the most part ignorant and hard to get along with, since they are not wise, but only appear wise.

Today the practice of rote memorization has been largely discarded in school curriculums, and we're content to outsource memory to our phones or computers. But some people still use the method of loci. The majority of memory athletes, for instance. These are people who compete in the World Memory Championships and hold incredible records for memorizing the precise order of binary numbers presented over five minutes (over 1,000 digits total), or random words over fifteen minutes (300). In 2002, the Irish neuroscientist Eleanor Maguire decided to investigate why some people have better memories than others. She used neuroimaging sensors to watch which neural mechanics were at work when memory athletes memorized information. Ten individuals deemed "superior memorizers" were tested against a control group; none of them demonstrated exceptional intellectual ability. The only difference Maguire found was the *part* of the brain they used to recall information. The neuroimaging sensors found that while the right cerebellum was basically active in everyone tested, the memorizers also showed activity in the left medial superior parietal gyrus, bilateral retrosplenial cortex, and right posterior hippocampus—many of the brain regions implicated in spatial memory and navigation.

In order to memorize a series of numbers, then faces, and then photographs of detailed snowflakes, nine out of the ten superior memorizers had utilized a route strategy to recall the information. When the new items were presented, they placed them in a familiar loci and then recalled them by revisiting those places later. "The longevity and success of the method of loci in particular may point to a natural human proclivity to use spatial context—and its instantiation in the right hippocampus—as one of the most effective means to learn and recall information," wrote Maguire and her coauthors in their study. As early as 1970, the Stanford psychologist Gordon Bower described the method of loci as a "journey" and a "mental walk" technique. The mnemonist creates a vivid mental place, akin to the brain's spatial representation of an actual place, and navigates it during the quest for a specific memory. The method of loci takes a piece of abstract, disembodied information and gives it a spatial organization that transforms it into a memory supported by the hippocampus.

While Maguire didn't find any structural brain differences between memory athletes who practiced the method of loci and a control group, she had previously found evidence that the hippocampus—the circuit in our brains responsible for building spatial representations used in navigation—is remarkably plastic. In 2000, she and a group of scientists at University College London published a study focused on the brains of London's taxi drivers. To get a license to drive one of the city's ubiquitous black cabs, drivers have to acquire what is called "The Knowledge," which includes memorizing some twenty-five thousand streets and thousands of landmarks. Maguire wanted to know if these drivers would have more gray matter, the tissue containing synapses and a high density of neuronal cell bodies (the nucleus-containing center of the neuron), in their hippocampus as a result of this knowledge. The researchers used magnetic resonance imaging (MRI) scans and found that, amazingly, the answer was yes. London's taxi drivers had significantly greater volume than a control group in their posterior hippocampus; it appeared that the number and complexity of navigational tasks a person practices influences the amount of gray matter.

Maybe, the researchers wondered, individuals with larger hippocampi were predisposed to become taxi drivers, a profession with a dependence on navigational skills. Yet their data showed that the amount of time spent in the profession correlated with greater volume, proving that the growth was accumulated. It was the environmental stimulus itself, the practice of navigation over time, that enabled plasticity, an ability to adapt and change, in this structure of the brain. Six years later, Maguire, Hugo Spiers, and Katherine Woollett published another study, comparing London's bus drivers to taxi drivers. Both were navigating the same city, presumably dealt with the same levels of stress, and possessed similar levels of driving experience. The difference between them was that whereas the taxi drivers had to take novel routes that changed day to day depending on their passengers, the bus drivers followed fixed routes. By comparing the hippocampal matter in these two groups, the researchers hoped to conclusively understand whether driving itself was responsible for more hippocampal volume in taxi drivers, or spatial knowledge. Again, the taxi drivers were found to have greater gray matter volume.

The malleability of the hippocampus is arguably one of its most important features. Might an individual be able to influence through practice, environment, and skill their own cognitive potential? It seems so. The more time spent learning and practicing, the greater the gray matter, scientists have discovered, in the various parts of the brains of musicians, bilinguals, even jugglers. "The results from the present study continue to permit the view that learning, representing, and using a spatial representation of a highly complex and large-scale environment is a primary function of the hippocampus in humans," the researchers reported, "such that this brain region might adapt structurally to accommodate its elaboration."

—

Before neuroscientists had discovered the role of the hippocampus—and its plastic qualities that could be harnessed for highly skilled navigation—the American aviator Harold Gatty argued unequivocally against the myth of a sixth sense, pointing instead to the seemingly limitless human capacity for learning. His book *Nature Is Your Guide* is a compendium of arcane and fascinating instructions about how people can learn to navigate without instruments. He wrote it in the late 1950s after an illustrious career teaching navigation on land and in the air, after he had retired to live in Fiji. (He died unexpectedly from a stroke just four months before it was published.) The book represents a lifetime of knowledge accumulated by someone who loved adventure and seemed to be entranced with solving the puzzles of orientation and piloting using logic and his senses. Indeed, Gatty believed that in the process of "evolution of our civilisation," people had lost what was once essential to our very survival: the power to observe nature.

Born in Tasmania, Gatty went to a naval academy at the age of fourteen and taught himself how to tell time by the stars while working on a steamship. He eventually opened a school of navigation in Los Angeles, teaching students how to read the sun and the stars. He invented a sextant to use in planes, and in 1931 he set a speed record by circumnavigating the globe in eight days. Gatty created Pan Am's first transpacific air service route; Howard Hughes called him a "trail-blazing pioneer."

Gatty's book encompassed the globe, from deserts, mountains, and polar regions to oceans. His strategies for navigating these topographies drew on history and knowledge of Native

Americans, Australian Aboriginals, Polynesians, Inuit, Europeans, and Saharan nomads. At every opportunity, Gatty emphasized that while those who grew up learning these skills will have "keener perceptions and more highly developed powers of observation than most of us," anyone was capable of developing their memory, sense of time and distance, and observational skills with practice, in order to become what he called "natural navigators." There was no special biological hardware at work in people who found their way without instruments. Their mastery was the result of tradition, a lifetime of exploration, and exhaustive knowledge of the landscape. If this was hard for his readers to grasp, Gatty said, that was because they had learned a conventional, Western version of history that taught them it was white explorers who had "discovered" native people in unexplored parts of the world. Gatty flipped that history: those same "natives," he pointed out, had discovered their homes in much earlier and unaided feats of navigation.

Gatty believed technology itself had perpetuated a myth of biological difference between races. "[Scientists] build a wall of mystery, fable and myth around the natural navigations of the past. So used are we to navigating by the compass, the chronometer, the sextant, the radio, radar and echo sounder, that some of us just cannot believe that early peoples could make long journeys into unknown areas, and find their way through unexplored wilds and across uncharted seas with only their normal senses and traditional wisdom to guide them."

These skills reminded Gatty of the Greek myth in which Ariadne gives Theseus a thread so he can find his way out of the maze of caves after slaying the Minotaur. In most places, said Gatty, the thread is an imaginary one. It is, just as Awa had described, a thread of memory.

WHY CHILDREN ARE AMNESIACS

The thread of memory that allows humans to explore without getting lost is one of humankind's most fascinating cognitive powers. But there is a period when memory fails each of us. In infancy and early childhood we experience the world and formulate episodic memories—our recollection of events and autobiography—only to have them disappear and become unreachable to us in adulthood. Until I learned this, I was confused about why my early childhood memories were so fugitive until the age of six. Before that, there's little I can grasp or distinguish as real versus imagined. After that, however, it's as though my memory is ignited. I remember we moved to a rural New England town of just a few hundred people and rented a trailer at the end of a dirt road, nestled between a cow field and a long wooden chicken coop in which I could crawl and gather eggs. My mom planted a garden of herbs, flowers, and vegetables next to the coop and lugged a splintering telephone pole to its middle, putting an old green pie dish on top to make a bird bath. She canned fruit and cooked in a heavy iron skillet. After dropping me off at school in an exhaust-spewing Chinook camper truck, she taught autistic adults living at a nearby sheep farm how to weave on giant clacking looms and worked as a waitress.

My dad, a housepainter, found work on the grander homes in the area or commuted to paint big Victorian houses. On the outside, our trailer was the quintessential redneck abode, but my parents were an odd amalgam; blue collar with serious spiritual aspirations, they counted their dimes to pay for pilgrimages to India and California.

What's curious to me is that to this day I can draw a perfect map of this happy place. I know exactly where the grapevines grew along the granite-stone-lined ditch, the distance to the gnarled pear tree, the beehives, and every individual young pine tree on the perimeter of the cow pasture. I still know the curve of the stream and the exact place where it swelled to create a muddy bathing hole for me and thousands of tadpoles in the early summer, and its meandering path into a thicket and beyond to a beaver pond. I can draw each apple and peach tree, blackberry and raspberry patch, towering birch, and the dirt path through the trees to the giant field of goldenrod. I remember the placement of stones inside a bramble of old lilac bushes where no one else except for me could fit. There I camped out, secluded in my own private world, the first room of my own.

I would wager that if I gave you this map and transported you back in time, its to-scale detail would allow you to navigate this place accurately. And remarkably, my memory extends beyond our home to the entire town. Why did my mind have such a frustrating void of my early years and the sudden lucidity and exactitude of my spatial memory after the age of six? For a while I chalked it up to some sort of trauma simply because the four years we lived in the trailer were the most stable of my then-young life and, it would turn out, the last secure moments of my adolescence. We left it for a period of repeated displacement and then divorce; from the age of ten until I was twenty-eight I would never live in the same place for more than a year or two. Did this period of stability explain the resolution in which my mind could recall our ramshackle Eden? Did I remember it so well simply because I had been happy? And why, I wondered, did my memory of these years so often take the form of a map that included the myriad routes and places where I had explored and run riot?

It was a neuroscientist who first told me about the universal phenomenon of amnesia in children. Kate Jeffery is an English neuroscientist whose laboratory at University College London studies the behavior of hippocampal cells in rats. At the core of her interest is the mystery of why the human brain seems to use the same neural circuit for navigating space and episodic memory; she has called it one of the most outstanding questions about the brain. "Why would nature have used the same structures for both space and memory, which seem so very different?" she wrote in *Current Biology*. "An intriguing possibility is that the cognitive map provides, in a manner of speaking, the stage upon which the drama of recollected life events is played out. By this account, it serves as the 'mind's eye' not only for remembering spaces, but also the events that happened there and even—according to recent human neuroimaging evidence—imagination."

I met Jeffery at a conference in London and asked her about my own experiences with childhood memory. Was there an age at which our cognitive powers for spatial mapping are "turned on," so to speak? How the spatial system develops in infancy, Jeffery told me, is still very much an open-ended question; we don't know how much of our brains are hardwired or how much spatial experience is necessary in order to condition the functions of the brain. Some studies have shown that animals raised in featureless or small confinements struggle with simple spatial tasks, but how this translates to humans is unclear. "I think the field is still grappling with these issues. We're not exactly sure. But there is a phenomenon, this period of time in infancy and beyond during which we don't have lasting episodic memories, we don't seem to be laying down those memories," she said. And, Jeffery pointed out, young infants don't form cognitive maps in the way adults do. "Their spatial organization of information is a lot less rich," said Jeffery. "It's possible that memories you form as an infant, because the hippocampus is still developing, may get overwritten or disturbed by the new circuitry that is still developing. And as an adult, you can't retrieve those early life memories the way you can later ones."

The hippocampus of rats is anatomically similar to humans, and for Jeffery, looking into their brains and listening to the activity of their neurons firing as they move provides tantalizing glimpses into the physiology of spatial mapping and memory. As we sat together talking about these questions, I asked Jeffery if she could explain the process by which neuroscientists think the hippocampus perceives and creates representations of space. Jeffery graciously took a piece of paper and pencil and began sketching a series of boxes and arrows, building a classic circuit diagram to illustrate the neural components of the hippocampus. She started with a box representing the entorhinal cortex, labeled it "EC," and split it into five layers representing various cell types. The entorhinal cortex, she told me, is the main interface between the neocortex, the part of the human brain associated with higher intelligence, and the hippocampus. All of the primary sensory areas—vision, olfaction, audition, touch, what Jeffery described as "a little bit of this, a little bit of that"—feed into the entorhinal cortex. From that box she began drawing arrows to other boxes labeled "DG," "CA3," "CA2," "CA1," and "SUB." These were the main components of the hippocampal circuit, each one fed by the various layers of the entorhinal cortex. "By the time you get to the hippocampus, quite a lot of stuff has happened, these senses are very highly processed," she explained. "But it turns out that layer two goes to CA3, layer three goes to CA1 and the subiculum, but there is an output from CA1 that goes back into layer five of the entorhinal cortex." She paused, looked at my furrowed brow, and chuckled. "So it's sort of like that but there's lots of backwards and forwards."

In recent years some of the most stunning images of the hippocampus have emerged from Harvard University's Center for Brain Science. There the neuroscientist Jeff Lichtman has pioneered a way of using microscopes to map neural connections in the brains of mice. By fiddling with genes, Lichtman causes mice to express different fluorescent proteins in individual neurons, which appear under magnification in bursts of beautiful pinks, blues, and greens. These "brainbow" photographs show how cells in the hippocampus are condensed into single orderly layers. Whereas neurons of the cortex look like a galaxy of randomly strewn stars, those in the hippocampus are aligned in elegant curving arcs.

These are the cells, called pyramidal neurons, that Jeffery and so many other neuroscientists are fascinated by. And they are a key to understanding the phenomenon of amnesia in our early lives.

Sigmund Freud coined the term "infant amnesia" and explained it in terms of repression; the brain was hiding the desires and emotions of infancy from the adult psyche, and these could be accessed through psychotherapy. "Hitherto it has not occurred to us to feel any astonishment at the fact of this amnesia, though we might have had good grounds for doing so," Freud wrote in 1910. "For we learn from other people that during these years, of which at a later date we retain nothing but a few unintelligible and fragmentary recollections, we reacted in a lively manner to impressions, that we were capable of expressing pain and joy in human fashion, that we gave evidence of love, jealousy, and other passionate feelings by which we were strongly moved at the time, and even that we gave utterance to remarks which were regarded by adults as good evidence of our possessing insight and the beginnings of a capacity for judgment. And of all this we, when we are grown up, have no knowledge on our own! Why should our memory lag so far behind the other activities of our minds?" Freud thought of memory as a permanent storage system that enacts a lasting influence over our behavior into adulthood, even if our conscious minds can't unlock it. What he didn't know is that this period of infant amnesia until the age of two—followed by childhood amnesia until the age of six or so—is not only universal among humans but some mammals as well. All altricial species who raise their young, including rats and monkeys, experience a period of amnesia, hinting at a potential evolutionarily conserved necessity for this developmental period.

From the 1970s to the 1990s, another explanation for infant amnesia was a child's lack of language: early memories become inaccessible once babies transition from nonverbal to verbal communication. It's precisely around the age of eighteen months that there is an explosion of language in infants, and shortly thereafter infant amnesia dissipates. As Nora Newcombe, founder of the Spatial Intelligence and Learning Center at Temple University, explained to me, "[They believed] that the advent of memories has to do with both language acquisition and is then tied up with cultural norms about the importance of remembering unique events. These are obviously not unimportant; we speak, we live in social groups. But that idea wasn't going to be enough. It wasn't going to be the only explanation." Further complicating the language hypothesis was the fact that so many animal species that never develop language nevertheless seem to remember events in their lives.

It's only more recently that scientists have uncovered connections between the development of spatial representation in children, amnesia, and memory, connections that might illuminate how these cognitive abilities evolved in humans in the first place. The ability of the human mind to engage in mental time travel—the ability to recall the past and imagine the future—and grammatical language may have evolved during the Pleistocene, the epoch that began 2.6 million years ago. This was also the period that it seems children, an Old English word that means "recently born," emerged as a new, prolonged stage of biological and social development in our species. As several researchers explain in the book *Predictions in the Brain*, edited by the neuroscientist Moshe

Bar, "This emergence of the genus *Homo* was accompanied by a prolongation of the period of development from infancy to adulthood, and that extra stage, known as childhood, was inserted into the sequence of developmental stages. Childhood lasts from 2½ to about age 7, roughly the period during which both mental time travel and grammatical language develop."

Are children the result of a novel evolutionary stage in our species' development, one that was needed in order for our brains to be able to fully develop our spatial and episodic memory systems? "Everyone thinks the first two years are so important, but if we can't remember them, *how* are they important?" said Newcombe. "There are some answers, but if we can't answer it crisply, that tells us we don't really understand anything about the brain."

When Jon was prematurely born at twenty-six weeks, he weighed around two pounds. He had trouble breathing on his own, and for the next two months he lived in an incubator and was hooked to a ventilator. But he grew into a healthy baby and toddler until the age of four, when he had two epileptic seizures. It was about a year later that his parents began to notice that Jon couldn't remember things that happened in his daily life. He didn't remember watching television or what had happened at school or what book he had read the night before. When a team of neuroscientists evaluated Jon, they discovered other impairments. He couldn't find his way anywhere, remember familiar environments, or locate objects or belongings. Remarkably, Jon's IQ was normal, he could read and write and spell, and he did well at school. His semantic memory, the recollection of facts unbound from personal experience, was intact.

For over a century, it has been the absence of memory in individuals like Jon that has given scientists a method for studying memory. Perhaps the most famous case of amnesia in the scientific literature is H.M., an epileptic who in the 1950s at the age of twenty-seven had part of his temporal lobes removed and lost his ability to acquire and recollect memories. H.M., whose real name was Henry Molaison, described his conscious existence as "like waking from a dream." His surroundings always seemed unfamiliar to him, he was forever in a "new" place. It took him many years to eventually memorize the floor plan of his own house. For these reasons, he could never remember the people who spent decades testing his memory or his way around places he had been visiting for years. One of these was the Massachusetts Institute of Technology (MIT), where Molaison was often a subject at the Behavioral Neuroscience Laboratory between 1962 and his death in 2008.

It was H.M.'s case that led scientists to initially identify the hippocampus as the source of episodic memory—the ability to formulate and recall the places and events that make up our autobiographical past. And in Jon's case, neuroscientists discovered the reason he couldn't remember his past or reliably find his way when they used magnetic resonance imaging to look at his brain. The lack of oxygen to his brain as an infant, known as hypoxia, and the subsequent seizures had caused rare and severe damage to the cells in his hippocampus, stunting its growth. As a result, it was abnormally small, about half the size of a healthy hippocampus. Jon was one of several children who became part of ongoing studies into the nature of hippocampal amnesia. "These kids are really incredible," Newcombe said. "There are only four or five of them, and their brain damage varies. But they are pretty normal: they go to school, they talk, they know facts, but they really don't remember their lives—they don't have autobiographical memory. And they can't find their way to a school that they have attended for years and that's two blocks away."

As it turns out, there are interesting parallels between amnesiacs like Jon and all children in the early years of life. Children's sense of space and memory for places and events is strange, far more emotionally vivid and sensitive than an adult's yet confounding because it is so fleeting and fragmented. They are capable of forming memories, but they are highly vulnerable to forgetting;

their memories are filamentlike and burn out quickly. Decades after the first scientific papers documenting H.M.'s condition and Jon's case were published, the scientific understanding of the hippocampus and its importance to childhood development and memory is rapidly growing. Scientists have found multiple types of cells in the hippocampal circuit: head-direction cells discharge in relation to the way our head is pointed on the horizontal plane, and grid cells fire as we roam an environment, building a coordinate system for navigating. Place cells fire at a unique location in space, what is called the place field. Though the presence of these cells in the human brain is often inferred with brain scans, researchers have proven they exist by recording their activity during epileptic therapies, when electrodes can be directly implanted in the brain. Other spatial cell types are distinct to certain taxa. For instance, monkeys (but not rodents) possess gaze-direction cells that fire when a monkey is simply looking in a particular location.

All together, many now believe it's the constellations of these cells blinking on and off in our brains that make self-localization and navigation possible. And there is evidence that infancy and toddlerhood are important periods in which cells in the hippocampus begin to encode space—as some say, map it—and mature. So, as babies explore their environment and create spatial representations, these experiences may be laying the neural foundation for episodic memory, our ability to remember the events of daily life.

The neuroscientist Lynn Nadel became interested in the developmental story of the hippocampus in the 1970s during the period he was researching and writing *The Hippocampus as Cognitive Map* with John O'Keefe, an eminent figure in the field of memory research. As they write, the hippocampus is a structure that matures at different times in different animals, unlike some other parts of the brain that are relatively mature at birth. In rats and mice, for instance, around 85 percent of the cells in the dentate gyrus—the sensory input region of the hippocampus—originate after birth in the days that correlate to the first two years of life for children. "The biggest surge in synaptic formation occurs in the period between postnatal days 4 and 11 when the number of synapses in the exposed blade doubles every day and the synaptic density increases 20 times."

They proposed a fascinating trigger for the spatial mapping system in the brain to begin creating these representations—exploration. Animals are engaged in mixtures of activities: nesting, foraging, walking, swimming, flying, sleeping. They also explore, a behavior that occurs when an animal encounters a place that is unfamiliar or novel and begins to gather information about it by physically investigating. In the context of the cognitive map theory, Nadel and O'Keefe said that exploration is critical for building the map, for cells to encode space and make what is unknown familiar. Novelty is when an item or place "does not have a representation in the locale system and thus excites the mismatch cells in that system." If the hippocampus disappeared, they predicted that exploratory behavior would also disappear in animals, and, in fact, lesion studies show this to be true. But what is the reason for this delayed maturation of the spatial mapping system? Perhaps it is to prevent young animals, who are still dependent on their mothers, from leaving their nest to explore and put themselves at risk.

After the book was published, Nadel continued to mull the implications of delayed maturation. "We had the theory of what the hippocampus did, but what does it mean if the hippocampus *doesn't* work?" Nadel told me. "What's it like if you don't have it? What's it like to be a relatively late-developing system? Does it make it more susceptible to plasticity in the environment?" Eventually he realized that the answer was amnesia. If the hippocampus wasn't working, according to their theory of the cognitive map and its support for memory, there would be an inability to remember anything. Nadel had inadvertently found a neurobiological explanation for infant amnesia: like Jon, we can't retain memories as children because we lack a fully functioning hippocampus.

In 1984 Nadel published his hypothesis, which he rooted in

the fact that the time during which children exhibit amnesia matches the postnatal maturation of the hippocampus in rats. With coauthor Stuart Zola-Morgan, he proposed that episodic memory is only possible after the brain is capable of place-learning and that infant amnesia is a period during which the hippocampal memory system for space is relatively undeveloped. Animals don't explore haphazardly, they pointed out, but in a structured fashion, going one place and then another but only rarely returning to locations they have already visited until they have sampled widely. "This pattern suggests the existence of internal representations that capture the spatial structure of the environment," they write. In rats, guinea pigs, and cats, exploratory behavior emerges just as the hippocampal system approaches maturity: "If the machinery is not there, the system will not function." And in both young animals and children, the capacity to store information about environments for spatial exploration and place-learning allows the encoding of events and where they occur, increasing memory capacity.

Three decades after he published the idea, Nadel told me that he thinks it is probably too simplistic—in both its definition of infant amnesia and the nature of development in the hippocampus, which varies between species. "The hippocampus, it's not like a structure that isn't there and then is there the next day. There is a gradual emergence of its functions," he said. "We now know more about the piecemeal nature of that. There's a patchy developmental picture that has emerged. On top of that, we know now that good episodic memory requires *more* than the hippocampus. For four or five years we don't have these memories, and it has more to do with the entire network, connections to prefrontal cortex and all the parts of the brain that are involved when we do fMRIs.* But the core idea that the first nine to eighteen months of life there is no episodic memory at all, that is still on the right

---

* Functional magnetic resonance imaging (fMRI) is a type of neuroimaging that detects changes in the brain's blood flow to infer brain activity.

track. It's the maturation of the network and the connections be-tween those parts that give you long-term memory."

Nadel and Zola-Morgan articulated a central mystery of spa-tial cognition: are we born with a brain that is hardwired to de-velop spatial memory or is experience important for building its infrastructure? Since then, hippocampal development and its rela-tionship to memory has remained one of the most intriguing is-sues in neuroscience. As Jeffery said, "People started to look at development, and there's quite a bit of interesting work now sug-gesting head-direction cells come online first, then place cells come online, and then the grid cells." Indeed the evidence sug-gests that while certain components of the cognitive map are in-nate to our brains, we undergo a period of acquiring spatial knowledge in early life that influences how well we perform these functions later on.

In 2010 two different research teams did something amazing: they implanted electrodes into freely moving preweaned rats the size of quail eggs and recorded individual neurons in the hippocam-pus. The teams, one at the Norwegian University of Science and Technology and the other at University College London, were able to record hundreds of head-direction cells, place cells, and grid cells starting from the rats' sixteenth day of life for two weeks. Both teams discovered that all three of the cell types were present in the young rats as early as two days after they opened their eyes, and *before* they began to leave the nest and explore their environ-ment. But of these cell types, only the head-direction cells were fully mature. It took several weeks of exploring the environment for the place cells and grid cells to become adultlike. From these data, the teams concluded that spatial learning continues to im-prove long after the components of the cognitive map are in place. Furthermore, one of the most important factors for determining the number and maturation of neurons was the age at which the young rats were exposed to new places, rather than how often they were exposed. The younger the age, the more quickly and easily they seemed able to encode spatial cells and learn.

Research into primates and behavioral studies of children have given neuroscientists clues to how the same process might occur in young humans. The Swiss neuroscientists Pierre Lavenex and Pamela Banta Lavenex have proposed that around two years of age the CA1 region of the hippocampus, essential to object differentiation in long-term memory, matures. Over the subsequent years of toddlerhood, the dentate gyrus, a remarkably plastic region of the brain that undergoes neurogenesis—the creation of new neurons—into adulthood, matures and supports the creation of new memories. By six, children show a strong positive relationship between hippocampal volume and their episodic memory—the bigger the volume, the greater the ability to recall details of an event—and six is the average age at which childhood amnesia diminishes.

Throughout this period, learning seems essential for the hippocampus to generate and condition neurons. Indeed, without sustained opportunities for children to experience what could be described as exploratory wayfinding, some researchers believe that there would be costs to cognition and memory. In 2016 researchers at New York University's Center for Neural Science published findings that showed how susceptible the development of the hippocampus is to learning experience. The team chose two different developmental ages in infant rats: postnatal day seventeen, which roughly corresponds to the age of two years in humans, and postnatal day twenty-four, which roughly corresponds to between six and ten years of age. By measuring molecular markers in the hippocampus, they were able to show how experience impacted the maturation of the hippocampus during this period. Then they increased or decreased the level of these molecules, thereby manipulating the rat hippocampus to either speed up memory retention or lengthen the window of infantile amnesia. What they concluded is that infantile amnesia is a type of critical period—a window of plasticity when environmental stimulation actively shapes the brain.

"Critical periods are when the system is particularly sensitive—if it doesn't receive the right stimuli, it is stunted," said Alessio Travaglia, a postdoctoral fellow and author of the study. "The brain is maturing through experience. We do think that without the right stimulation the hippocampus is not going to develop, actually. It's not just the infantile amnesia, the fact that there is a critical period in maturation we think has big implications for education and what kids need." He used the example of an eye. "This was an early experiment done in the sixties. If you put a patch on your eye and keep it closed for a week, after a week you can still be fine. If you close the eye of a young animal early in life during this critical period, they found the animal cannot see, it's going to be blind. Another example of a critical period is language. For example, if a baby learns another language when it's very young, it's going to be fluent."

Travaglia and his fellow researchers think the hippocampus needs experience and opportunities to mature. "For humans, the assumption is that the brain needs the right stimulation in this critical period too. The right stimulation, meaning that kids should experience the right noise, games, environment, playing, and if they lack these stimuli they might have an effect later on," he told me.

One possible developmental milestone is when a child transitions from being passively carried to self-mobile. Perhaps it's this change in movement that impacts how spatial information is encoded in memory? In 2007, for example, a group of researchers in England found that the onset of crawling in nine-month-old babies was associated with a cognitive leap: a more flexible and sophisticated capacity for memory retrieval. Arthur Glenberg, a professor of psychology at Arizona State University, has a hypothesis that the onset of self-locomotion prompts hippocampal maturation: once babies begin moving through space on their own, their place cells and grid cells can begin aligning themselves to the environment, ultimately facilitating the creation of the infrastructure of long-term memory. He thinks the tuning of these cells depends on the consistent correlation between optic flow, head direction, and the unconscious perception of spatial orientation from self-generated movements; until infants begin moving themselves through space, the whole system is immature and therefore an unreliable contributor to memory. When infants begin to crawl and explore space, conditioning the spatial location coding, that movement can become the scaffolding for long-term episodic memory, and forgetting declines. Glenberg's hypothesis also provides an intriguing explanation for memory degradation in older age: as the body ages there may be less self-locomotion and exploration. Perhaps, he suggests, hippocampal place and grid-cell firing become dissociated from the environment and result in less powerful memory recall.

Glenberg's idea doesn't fully explain why there is such a large gap between the *start* of self-locomotion in the first year of human life and the reliable retention of memories that takes place around age six. He suggested that the tuning of the hippocampus to the environment by crawling has to be relearned once we begin walking. But it's also possible that this gap comes down to experience and how much is required. It takes time to explore enough space and begin formulating complex cognitive maps and a fully functioning hippocampal memory system anywhere near the sophistication of an adult's. In fact, the age of self-locomotion seems to matter less than the amount of exploration children engage in. Dutch researchers found in 2014 that by the age of four, children who spent more time exploring had a higher capacity for spatial memory and a positive correlation to fluid intelligence—solving problems, identifying patterns, and logic. "Your ten-month-old knows his way around the apartment but would not have much luck getting from the apartment to the park," Glenberg told me. "It takes an awful lot of experience walking around to develop this complex-enough set of cells that can serve as a good substrate for memory."

In 1999, a group of researchers led by Rusty Gage at the Salk Institute for Biological Studies in California discovered that exercise induces neurogenesis in the adult hippocampus, specifically the dentate gyrus, the region through which the hippocampus receives most of its connections from other parts of the brain and which is implicated in forming episodic memories. More recently, three researchers at the National Institutes of Health's program on aging looked at the brain cells of adult mice that spent a month with running wheels in their cages, a group that had wheels for one week, and a group that had no wheels. Afterward, both groups with wheels had brains that had developed new neurons and longer dendrites compared to mice that had no wheels. Running, the researchers concluded, likely facilitates spatial information encoding by increasing the generation of neurons while also reorganizing neuron circuitry.

The fact that hippocampal development is influenced by these kinds of activities and experiences indicates an incredible plasticity, with implications for childcare, education, and the treatment of cognitive impairments. "It's very exciting because the maturation of the brain is often considered dependent on time and a genetic program," Travaglia told me. "What we're showing is that the development of the brain is not a fixed program, it's about *experience*."

In the 1940s the psychologists Jean Piaget and Bärbel Inhelder tested children on a "three mountain task." They placed a doll on different parts of a small-scale model of three mountains and asked the children to select one of several pictures to match the doll's perspective at each spot. At four years of age, most of the children couldn't distinguish their point of view from the doll's, leading the psychologists to believe that young children rely on the more elementary egocentric perspective that precedes logical thinking. Later, around nine or ten years of age, children, they thought, switch to an allocentric representation, the ability to encode the Euclidean, objective relationships between landmarks and assume the perspective of multiple objects to each other.

Later research has shown this classic development sequence from egocentric to allocentric in children to be flawed: Newcombe has shown that babies as young as twenty-one months can accurately represent locations allocentrically. In a 2010 study published in the *Journal of Experimental Child Psychology*, Norwegian and French psychologists tested seventy-seven children in elementary school using a virtual maze task. They found that while all of the five-, seven-, and ten-year-olds used a sequential egocentric strategy to solve the task, they were able to adopt an allocentric strategy too. Even the youngest chidren tested could do it. But, the older the children were, the more spontaneously they could transition to the allocentric perspective and use it with greater accuracy; ten-year-olds were able to orient themselves at the beginning of the task and create an abstract top-view representation of the maze equal to that created by adults.

The findings suggest that while young children are able to employ the allocentric strategy, the nature of it changes progressively between five and ten years of age. By ten, individual children can demonstrate startling variations with their peers in hippocampal volume. Researchers have found that children who have higher levels of physical fitness have larger hippocampal sizes than those who are less active, indicating a relationship between aerobic exercise and the structure of preadolescent brains. Furthermore, those structural differences seem to impact function. The same ten-year-olds who were more physically active and fit showed better performances on memory tasks.

We're not the only animals that demonstrate the plastic nature of the hippocampus and its relationship to cognitive ability. In nonhuman primates, hippocampal volume is the measure most consistently correlated with performance on spatial and nonspatial tasks, and can even predict performance. University of Oxford researchers Susanne Shultz and Robin Dunbar have looked at forty-six different species of primates, including gorillas, lemurs, and macaques, and given individuals eight different tasks designed to test learning, memory, and spatial cognition. Those primate species with a larger hippocampus performed better. Proportional brain size in primates has been found to correlate with social learning and tool use, the formation of coalitions, the ability to deceive, and the size of social groups, all aspects of higher-order cognition or what is also called executive functions—the ability to organize thoughts and actions and direct oneself toward obtaining goals. The demand for increasingly sophisticated executive functions among primates may be one of the selective pressures that actually led to their brain enlargement (and, eventually, us).

Shultz and Dunbar have also discovered that birds that store food in different locations, sometimes returning days or months later to retrieve it, have a bigger hippocampal homologue. In one of their earlier studies in the late 1980s, they chose thirty-five different species and subspecies of passerine birds, an order that encompasses over half of all bird species who use their toes to perch, and dissected the brains of fifty-two specimens taken from the wild. Some of the birds belonged to species known to store food, and some relied on foraging alone. The researchers wanted to know: did food-storage techniques place greater demands on memory? And did birds that used these strategies develop special memory capacities that might be reflected in their brain volume? Shultz and Dunbar found that a bird like the marsh tit, which stores food in woods, had 31 percent higher volume in its hippocampus than a closely related bird, like the great tit, that forages.

Seven years later, Shultz and Dunbar decided to look at a single species, the common khaki-colored garden warbler. Would garden warblers with more migration experience have a larger hippocampus than those that didn't? If so, perhaps they were not unlike those taxi drivers whose memorization of the streets of London leads to a larger volume of hippocampal gray matter. They compared the brains of young birds that hadn't undertaken the annual migration from Europe to Africa to those that had, and found that birds with more migratory experience had significantly larger hippocampi—the result of both greater age and more experience. Other studies with pigeons have shown how their hippocampus is important for learning landmarks, and if researchers give them lesions in this part of the brain, they lose their ability to home.

Black-capped chickadees will not only return to places to retrieve hidden stocks, but they'll visit the stocks that hold their favorite food first and their least favorite last. This memory feat is surpassed by the humble scrub jay, which can remember not only where events happened but also when. The birds' favorite food is wax worms, but only when fresh; once they dry up, the worms are less delectable. Two researchers, Nicola Clayton and Anthony Dickinson, conducted a study in which they gave scrub jays wax worms to hide and, four hours later, they gave the jays the choice between retrieving the hidden worms or peanuts. But in some cases the researchers waited five days after they hid the worms to give the birds a choice. After four hours the jays chose to retrieve worms, but after five days they chose peanuts. They not only remembered what they had hidden but *when* they had done it. So do scrub jays have episodic memory?

The difference between humans and other animals and our cognitive abilities seems to have less to do with size and more to do with the sheer number of neurons that we develop and, crucially, *where* in the brain these neurons are located. An African elephant's brain is three times larger than ours and has three times the number of neurons. But its hippocampus has less than 36 million neurons compared to the 250 million neurons in ours. Nonetheless, some African elephants have been known to inhabit home ranges of over twelve thousand square miles; what specialized coordination of spatial memory and senses allows them to navigate these spaces? Some have surmised that they must use a hippocampal-dependent spatial strategy similar to humans. Meanwhile, whales also travel thousands of miles but have unusually small hippocampi and no detectable adult neurogenesis.

Pondering how animals experience the world stretches our imagination. The scientist Jakob von Uexküll thought that the behavior of an animal could only be explained by considering the inner, sensory world it inhabits. According to him, organisms inhabit their own *Umwelt*, a German word that means "environment," and he used this concept to explain how animals' subjective sensory experiences evolved to meet their needs. According to this idea, bees live in an ultraviolet world, because it allows them to orient themselves by polarized light, and wolves inhabit a landscape of smell in order to create landmarks and maps of places of import. Maybe the indigo bunting is blind to stars of lesser magnitudes in order to worship the North Star, its compass.

When it came to conceptualizing the intertwining relationships between organisms and their environment, Uexküll turned to the metaphor of music. Every organism is like a melody that resonates and harmonizes with living things around it. As he wrote, "All living beings have their origin in a duet." For children, that duet seems to be the interaction between the neurons firing in their mind and the places where they grow up.

BIRDS, BEES, WOLVES, AND WHALES

One morning in the Arctic I woke early and pulled on thick, windproof pants and an anorak with a hood edged in wolf fur and climbed over still-sleeping adults and children spread across the floor of a one-room cabin. I opened the plywood door a crack and snuck through it so as not to let the freezing air inside. Shoving my feet into heavy, felt-lined boots, I stood up and took in the sight before me. The cabin sat high on a hillside at the mouth of a large inlet covered in aquamarine sea ice that had been pushed by powerful tides into the coastline to form giant ruches. To get to this remote cabin we had sledded south from Iqaluit for hours, hugging an edge of the bay, past a place called Pitsiulaaqsit, which means the island where guillemots nest in Inuktitut, and another called Qaaqtalik, meaning the place where a mattress of caribou skins was left long ago. We had turned inland at Nuluarjuk, the island shaped like small buttocks. After spiking a long chain onto the ice and tying the dogs to it, we collected blocks of frozen freshwater from a nearby pond up in the hills, cutting them away with heavy metal spades and melting them to drink. For dinner we ate aged caribou ribs, delicate pieces of raw Arctic char, ptarmigan seared in its own blood, and boiled musk ox. Below I could see the teams of dogs where they had slept the night; they barely lifted their tucked noses to acknowledge my presence, and beyond them I saw the bay, a frigid expanse of white. The floe edge, where the ice meets open ocean, was still several hundred miles south of us.

I'd arrived on Baffin Island expecting most hunters to get around with dogsleds, but I quickly realized this was akin to showing up in New York City and wondering where all the horse-pulled carts were. While Greenlandic hunters are required by law to use sled dogs for hunting, and some remote communities in Nunavut still maintain sled dog teams for racing, in the whole of Iqaluit there were only about half a dozen teams. Everyone else used snowmobiles. I'd managed to ride to the cabin with a team of dogs belonging to Matty McNair, an explorer who had captained the first all-female expedition to the North Pole and has lived in Iqaluit for decades. Her dogs were well conditioned and had traveled all over Baffin Island, often on trips where McNair intentionally navigates using mainly landmarks, stars, and snow. But her dogs, she told me, are far better at finding their way than she is. "I don't know how they navigate: I've had dogs hit the town dead on in weather I couldn't even see in," she said. "It wasn't smell, you couldn't see anything visually, there were no trails they were on. It's just absolutely uncanny how they navigate. I've also been out at the beginning of the year and going on a snowmobile trail, and the dogs will turn off and go around the rock because that's where the trail went last year. They don't care where the snowmobile trail goes, that's where the trail went last year."

The Inuit sled dog is a distinct breed that made life possible in the Arctic for generations. Without them, travel on snow and ice used to be impossible; in the summer and fall they carried food and supplies across the uneven tundra. Dogs were so important that they were often fed first, then children, then adults. Today snowmobiles have undeniable advantages over dogs, which require nearly year-round hunting in order to feed them. As one dog team owner explained to me, to maintain a team of nine dogs, he has to provide four and a half tons of walrus and seal meat each year. For most hunters, the time required to feed dogs is simply too much of a burden. As the dogsledder Ken MacRury told me, "Inuit are pragmatists, they are not romantics. If the dog team is not useful, they're gone. Snowmobiles made it possible for people to keep full-time jobs and still be hunters. You didn't have to feed a snowmobile all summer."

The differences between riding a snowmobile and a dogsled are obvious. The former is much, much faster. But the slow speed of sledding provides an ideal pace for teaching and learning geographical and environmental knowledge, committing to memory landmarks, details of routes, place-names, and vistas. "The faster you traverse the land, the less observant of it you become," explained John MacDonald, a resident of Igloolik for twenty-five years who worked closely with the community's oral history project and is the author of *The Arctic Sky*. He once traveled with an elder in Igloolik who stopped at a rock and recognized it by the pattern of lichen on its surface. "I could have passed it and not even looked twice," said MacDonald, "which is exactly what you tend to do with snowmobiles." Snowmobiles also create an experience of always driving into wind, whereas hunters on dog teams travel slowly enough that they can use wind direction as an orientation tool to keep a bearing. Indeed, hunters across the eastern Arctic often used a wind compass with *Uangnaq* and *Nigiq* as the axis, and had up to sixteen terms to describe the in-between bearings. Often the Inuit dogs themselves played an important role in how people navigated, as McNair had described to me. A good driver (never called a "musher" in the eastern Arctic) rarely used a whip, if at all. The optimal relationship between the driver and his team is based on the concept of *isuma,* an Inuktitut word that means something like "mind" or "thinking," and in certain contexts it can mean "life force." The driver guides and directs his team using his mind, focusing his will on the team. The lead dog is the *isumataq*, which means "the one who thinks," and is the most responsive to the will of the driver. "You have to direct your *isuma*," explained MacRury. "You have to project your thoughts onto the dogs and they have to respond to that. . . . [Y]ou communicate with your dogs with your mind and your voice."

According to MacRury, it wasn't until his apprenticeship among hunters ended and he began running his own dogs that he started to realize how critical Inuit dogs were to finding one's way. Again and again he discovered that some of his dogs had uncanny abilities to navigate in any condition. "I had some wild experiences in total blizzard conditions, and the dogs would get you home. They seem to have a sense that we're going home and it's in this direction and they follow it. I'm sure they never deviated five feet from the trail, and yet I couldn't see a thing." Not all his dogs had this capacity; some were better than others. "They are not all cookie-cutter types, they have very distinct abilities," he explained. But the Inuit have been ruthlessly weeding out dogs from the breeding pool for hundreds of generations, producing what MacRury called simply a "pretty amazing animal." He trusted their memory to such a great extent that in blizzard conditions he stopped directing his dogs and simply let them go, confident that they would take him home even if there was no trail on which to backtrack. Indeed, his lead dogs were often able to take shortcuts in the dark across miles of unknown country until they picked up a main trail leading into Iqaluit. The feats MacRury described seemed to rely on the creation and retention of incredibly specific and detailed cognitive maps. But do dogs have them? John MacDonald told me that around Igloolik there is a word for this capacity to unfailingly know where one is regardless of external conditions, and that it can apply to both people and dogs: *aangaittuq*. "The term can be translated as 'ultra-observant,'" explained MacDonald. Its opposite is *aangajuq*, a term for "one who moves away from the community and immediately loses where his destination is at, so as a result will travel blindly."

In the 1970s a behavioral psychologist at the University of

Michigan argued that wolves do have cognitive maps. Roger
Peters spent several years observing wolves in the wild and came
to believe that they could create maps with a level of skill not
usually granted to nonhuman animals. Furthermore, the shared
capacity for cognitive mapping between men and wolves was
not a coincidence. Both species evolved as social hunters of big
game, meaning they formed groups and traveled over large areas
of space in pursuit of prey, then returned to their young, pack, or
camp. He estimated that the range of wolves and men was about
the same: both could travel around a hundred miles in twenty-
four hours. "Men and wolves have both had millions of years to
evolve solutions to the problems of getting lost, where 'lost' means
separation from your fellow-hunters, not knowing a quick way
back to your young, or where your prey is headed." Peters didn't
think this was a map in the sense of an aerial view but a simplification of the environment insofar as the brain threw away unnecessary information and retained and organized other elements, such
as the locations of dens, feeding grounds, water, food caches,
shortcuts, and predators and the spatial relationships between
them. For wolves, these maps were particularly dependent on olfactory cues, which Peters noted were a much more important and
vivid part of a wolf's world than humans could imagine. "For
wolves, the reality of an object may lie much more in its smell
than in its visual properties," he wrote. Peters knew from his
field research that wolves marked their path every three hundred meters on average, and particular attention was paid to
junctions—the intersection of paths that would most often serve
as rendezvous sites with others in the pack. The wolves were creating nodes, turning an otherwise blank landscape into a network
of landmarks.

———

Though it was early morning, the sun had already been up for
eight hours, showering the land in fluorescence. It was cold enough
that I pulled my arms in from the sleeves of the anorak to hug my-

self, and in this compact form I began walking up the rocky hill-side behind the cabin, sinking in pockets of snow and stepping over foot-high willow trees like a giant in a forest. Small as the trees were, some were likely over a hundred years old; most grow just a tenth of a millimeter a year in the Arctic. I was hoping to catch sight of snow geese arriving at this spit of land after a three-thousand-mile journey along an invisible avian superhighway with five hundred other bird species, each getting off at their respective feeding, breeding, and nesting places.

Inuit taxonomy classifies life under three categories: *anirniliit*, those that breathe; *nunarait*, things that grow; and *uumajuit*, which includes everything that moves. Some *uumajuit* are *tingmiat*, those that fly, and others are *pisuktiit*, the walkers. This last category is where humans belong, alongside caribou and musk ox. Snow geese are *tingmiat* and precious to hunters, who anticipate their arrival for weeks each spring. In Iqaluit I saw twelve-year-old kids with shotguns strapped across their chests riding snowmobiles off into the hills to look for them. Two young hunters, intoxicated with the possibility of shooting geese, told me how they had recently driven 120 miles on their snowmobiles in the hopes of finding some. When the geese start to arrive en masse, hunters can easily shoot sixty or more birds over a day to eat, freeze, and share. Now I searched for them in the silent moonscape, moving to stay warm.

Life on earth has created millions of Ulyssean species undertaking epic journeys at scales both large and small. Getting lost is a uniquely human problem. Many animals are incredible navigators, capable of undertaking journeys that far eclipse our individual abilities. The greatest migration on earth belongs to the Arctic tern, a four-ounce argonaut that travels each year from Greenland to Antarctica and back again, a distance of some forty-four thousand miles. Flying with the wind, the tern's return itinerary is a globe-trotter's fantasy, circumnavigating Africa and South America. Sooty shearwaters fly over thirty-nine thousand miles, zipping around the Pacific Ocean in figure eights to take advantage of prevalent winds. The ornithologist Peter Berthold estimates that half of all known bird species—fifty billion birds in all—migrate each year. Epic journeys are not limited to the feathered tribes. Undulating herds of zebras and wildebeests travel across the Serengeti to follow the rain. Leatherback turtles leave the coast of California and swim to Indonesia ten thousand miles away, and then make their way back again to the same beaches where they were born.

The lesser-known journeys are no less spectacular. The word *plankton* was coined by a German physiologist from the Greek *plazesthai,* meaning to wander or drift, and describes the tiny microorganisms that are carried by the ever-moving mass of the ocean. But plankton's random perambulations are only horizontal. Every twenty-four hours, trillions of these organisms, billions of tons of biomass in total, undertake an intentional *vertical* migration, rising to the surface of the ocean at twilight and descending at sunrise. Are these plankton similar to the first organisms that moved? Not the first to be swayed, pushed, flung, or caught by air or water, but the first to move from one place to another by their own volition? The earliest vertebrates, according to the authors of *The Evolution of Memory Systems*, developed a homologue to the hippocampus, giving them a navigation system that worked in tandem with older reinforcement systems. As they put it, this system guided behavior by linking stimuli and actions to biological costs and benefits, and just about all behavior in these ancient ancestors involved navigation: foraging, predator avoidance, temperature regulation, and reproduction. In order to survive, animals couldn't just move randomly but had to wayfind from one specific place to another, a demand that has resulted in a diversity of navigation mechanisms in nature.

Scientists have conceptualized this diversity as evolution's navigational toolbox. This idea was presented in 2011 by ten prominent scientists, including Kate Jeffery and Nora Newcombe, who study both animal and human cognition and behavior in the hopes of formulating common underlying principles of navigation. The scientists have broken the known mechanisms into four levels, from simple to complex. The first is the sensorimotor toolbox, which includes vision, audition, olfaction, touch, magnetism, and proprioception. At the second level they put "spatial primitives," animals that orient using simple representations and landmarks, terrain slope, compass headings, boundaries, posture, speed, or acceleration. At level three are more complex integrations of these tools to build spatial constructs, tools like an internal cognitive map. And at the fourth level they put spatial symbols: external maps, signage, and human language, essentially the ability to communicate spatial information. According to this idea, the simplest tools are fundamental—they appeared early in evolution and have persisted through the eons, and the more complex tools are synthesized from the early ones.

But conceptualizing animal navigation as a toolbox creates its own confounding questions. More often than not, animals once thought by scientists to use relatively elementary tools are discovered to have far more flexible and sophisticated tools at their disposal. Some animals seem to have all the tools, while others that would logically seem to require the most complicated tools make do with very simple ones. And some of the simplest tools in the box are ones we don't understand at all. We have evidence that they exist, but we can't see them and barely know how they work. For these reasons, the field of animal navigation today still contains some of the most fascinating biological puzzles in science. We have amassed a body of data based on tens of thousands of observations of animals navigating across the planet, yet we still grope to explain how they do it.

One of the devices that an animal needs to navigate is a "clock"— an internal mechanism for measuring or keeping time. The daily mass migration of zooplankton in the world's oceans requires them to know when dawn and dusk are approaching. It would seem this is a simple response to light stimuli, but deep-sea zooplankton, which live at depths below where light penetrates, also migrate in accordance with the length of day at different latitudes. Even slightly more complex migrations can demand multiple clocks. In their book *Nature's Compass: The Mystery of Animal Navigation*, James Gould and Carol Grant Gould describe the "eerily consistent" migration of Bermuda fireworms, a bioluminescent marine species that emerges en masse once every lunar month in the summer or, more specifically, fifty-seven minutes after sunset on the third evening after the full moon. To pull it off, the Goulds hypothesize that the fireworms must have a 27.3-day lunar clock, a twenty-four-hour daily clock, and an interval timer in order to measure the fifty-seven minutes from sunset. Animals that complete annual migrations or multiyear migrations have to possess a yearly clock, one that is finely attuned to the lengths of days and nights and their changes across each season. In all, evolution seems to have produced annual clocks, lunar clocks, tidal clocks, circadian clocks, and, perhaps for those that migrate under cover of darkness, a sidereal clock—which measures the time it takes a star to appear to travel around the earth.

One of the first people to discover the use of clocks by animals for navigation was an amateur entomologist fascinated by desert ants. Felix Santschi was a Swiss physician who left his home in Lausanne in 1901 and moved to a remote city in Tunisia. Santschi described and named almost two thousand ant species in his lifetime and was intrigued by their behavior, particularly how the ants outside the ramparts of the city he lived in navigated the desert. At that time, as the German neuroethologist Rüdiger Wehner has written, some surmised that ants oriented by scent trails; they foraged in one direction and then followed the scent they had laid down back again. But the desert is an environment where wind and sand create an ever-changing landscape, blowing away scents and landmarks that ants might use for orientation.

Santschi was the first to notice that ants weren't just following trails back and forth on their foraging trips, they were traveling in circuitous routes and then taking a direct route back. The ability to calculate a shortcut meant the ants were essentially doing trigonometry, calculating the spatial relationships among all the places they had visited and divining the straightest route home. Santschi knew this required a directional cue of some sort by which they could reliably orient themselves in space, and his guess was that ants were using a celestial compass, most likely the sun, by tracking its position at sunrise and throughout the day. To test this, he took a mirror to deflect the sunlight and observed that the ants changed their course back home by 180 degrees.

But the earth is not stationary, and the sun's position in the sky changes. For it to be an accurate navigational aid, an animal must change its angle of orientation over the course of the day to maintain a constant direction. Thus, Santschi thought ants had to have an internal representation of time in addition to the sun in order to derive accurate directions. Later he even tried limiting the ants' view of the sun entirely, and found they could still find their way based on even a small patch of sky. Subsequent biologists discovered that an ant's ocelli, light-sensitive photoreceptor organs on the head, can read information from a blue sky even when the sun and landmarks are obscured, using the polarized pattern of light to orient the ant and help it find its way home. It is, the entomologist Hugh Dingle ventures, a kind of "preprogramed 'hard-wired' representation of a celestial map."

Honeybees can also use polarized light to find their way. They have been called the most elegant of nature's navigators, embarking on up to five hundred trips a day, as far as five miles from their hives, in search of flowers and food. Like desert ants, they not only take circuitous, meandering routes in search of pollen but are always able to take the straightest route back home—a "beeline." How they manage to calculate shortcuts has been the subject of entire books and countless research papers—even Aristotle puzzled over them. Their feats are even more impressive because bees embark on far-reaching rambles with what we would deem considerable impairments. Their brains weigh less than a milligram and contain fewer than one million neurons, and they are blind by human standards with merely 20/2000 vision.

The biologist James Gould at Princeton University has been studying bee navigation for decades. On its surface, beelining seems to require what is called path integration, dead reckoning, or inertial navigation; by keeping track of each stage of a journey, the insects can compute their location and the direction home. But as a young biologist, Gould found that no matter where he displaced bees within their foraging areas, they were always able to find new shortcuts, suggesting that they have a flexible memory or internal representation of space. In other words, bees are using a far more complex evolutionary tool, what is often called the cognitive map. Bees not only seem to have an internal representation of space but appear to possess the ability to communicate this "map" to other bees, a capacity that, according to the navigational toolbox idea, is assumed to be specific to the human species.

In the 1940s, the Austrian scientist Karl von Frisch observed that bees, after discovering a rich source of food on their foraging journeys, flew back to the hive and began waggling their bodies in a figure-eight pattern. There was a very specific grammar to this dance, particularly if the bees were returning from flights greater than fifty meters away. The bees moved their bodies along a vertical sheet of honeycomb, and it turned out that the *angle* at which they positioned themselves against this surface to waggle was the angle their hive mates should fly in relationship to the sun. Furthermore, the duration of the dance was proportional to the distance of the food from the hive. The bees were giving directions to their hive mates to follow but using their bodies to illustrate the journey. Some bees would keep up the dance for hours, or restart the dance the next day, or even after several months of freezing weather, and their accuracy never seemed to suffer.

As he describes in his 1950 book *Bees*, von Frisch also discovered that like ants, bees were orienting in space using the sun as a compass, which meant they were also using internal clocks: a twenty-four-hour internal clock and a seasonal calendar tracking the passage of time. As to how honeybees learn the movement of the sun and its coinciding times, studies show that the third week of a honeybee's life is spent close to the hive, embarking on short flights, during which the bee learns the azimuth angles of the sun, its movement, and how to orient, before launching into long-distance foraging. In 2005 a team of German and British scientists led by Randolf Menzel described these early flights as a period in which young bees are forming an exploratory memory, perhaps the very cognitive map suggested by Gould. But they discovered that the map is far richer and more flexible than previously understood. In the journal *Proceedings of the National Academy of Sciences (PNAS)*, the scientists reported how they took three different groups of bees and displaced them in the night, tracking their flight paths with a harmonic radar—transponder antennas that were attached to individual bees and that emitted a wavelength picked up by a receiver. The bees recognized familiar landmarks from different angles and created new courses from arbitrary locations.

Monarch butterflies, lizards, shrimp, lobsters, cuttlefish, crickets, and rainbow trout as well as numerous migratory birds have been proven to use polarized light as a "compass," raising the question of whether this is a case of convergent evolution (a kind of coincidence of natural selection among independent organisms) or a shared ancient mechanism, present in the earliest species and carried through the eons.

While some animals navigate using the sun, others follow the stars. African ball-rolling dung beetles were known to navigate by the sun and moon, but in 2012 some scientists were puzzled by their ability to find their way even on moonless nights. They decided to release beetles with their dung balls in a walled space that limited their visual cues to the night sky and filmed their movements. It was clear that the beetles were orienting without the moon, meaning they must be using the stars to find their way. But how? The vast majority of stars, as the scientists pointed out, should be too dim for the beetles' eyes to detect. By taking the beetles into a planetarium, they soon discovered that the insects relied on the bright light of the Milky Way to get home. Likewise, southern cricket frogs, Namibian desert spiders, and large yellow underwing moths have all been proven to orient by the stars. Several bird species, including indigo buntings, European pied fly-catchers, and blackcaps, seem to orient by the North Star, using it as a rotational center to guide them through the night.

Even when we know the mechanisms animals use, the precision with which they use them often defies current scientific understanding. For instance, the clocks of many species are far more precise than the biological clocks available to humans. After just twenty-four hours of continual darkness, the human circadian clock is skewed on average sixty minutes from the actual time. For honeybees, this sort of inaccuracy could prove disastrous. Just fifteen minutes off, explains Gould, and a honeybee could have a ten-degree error in orientation, which over shorter distances might result in a couple dozen feet off course. And for long-distance migratory birds like bar-tailed godwits, errors in accuracy are deadly. Every autumn these birds leave their nesting grounds in coastal Alaska and head south to feed in warmer parts. The sensible route is along the continental arc of Asia to the eastern coast of Australia; there are plenty of landmarks and places to rest along the way. Instead, the godwits set out for the vast open water of the Pacific Ocean. For eight days and nights they fly over more than six thousand miles of featureless ocean before arriving in New Zealand. If the birds make a directional error of even a few degrees, they will be hundreds of miles off course, nowhere near their feeding and nesting grounds, and out of fuel.

Every year, humpback whales undertake migrations exceeding ten thousand miles across open ocean. These forty-ton mammals don't travel in the general direction of north to south and back again; instead, the whales return to the exact places they were born and fed by their mothers as calves, which requires exceptional navigation.

Recently a consortium of researchers led by Travis Horton at the University of Canterbury revealed just how precise whales have to be by tagging sixteen humpbacks and tracking their movements with satellite telemetry over seven years. What they found was that for most of the journey, the whales maintained a constant course that barely deviated even a single degree. This takes some unpacking to understand. Humans can maintain a constant course, but only if landmarks are available by which to judge our progress and correct our course as we go. Without such course corrections, we start unwittingly walking in circles. At the Max Planck Institute for Biological Cybernetics, for example, researchers Jan Souman and Marc Ernst have found that this predilection for circles became especially strong in blindfolded individuals; with their eyes covered, people quickly started walking in circles measuring sixty-six feet in diameter; this happens even when people *thought* they were walking in a straight line. Humpbacks follow a direction "as straight as an arrow," not just in short bursts but across thousands of miles. And they do it despite encountering multiple elements that could complicate their sense of direction. Humpbacks swim through storms, strong sea currents, and highly variable depth conditions; encounter entire underwater mountain ranges; and do so both during the day and night—and their course often varies less than one degree.

Humpbacks must be constantly compensating for these forces using spatial reference frames and orientation cues, but what are they? Perhaps humpbacks, like so many other species, are using a sun compass. Yet the researchers found that even when individual humpbacks began their journeys in different parts of the ocean, where the sun appeared at different altitudes and azimuths, they followed very similar headings. Other whales starting out in the *same* locations, where the sun appeared in the same place, used very different headings, meaning there had to be another reference they used to know their location. If humpbacks can't logically depend on the sun alone, what is the other reference? The navigation of so many species prompts this same question. While the sun, stars, moon, landmarks, olfactory cues, memory, and genetics seem to have a place in various species' strategies, none can fully explain how so many animals possess a powerful ability to orient with such remarkable exactitude. As a result, scientists have increasingly turned their attention to one of the senses at the so-called simplest level of the navigational toolbox to try and explain the most complicated migrations carried out by the likes of humpbacks and godwits: magnetism.

For decades, the idea of animal magnetic navigation was disparaged by the scientific community as pseudoscience. Then in 1958 a young graduate student in Germany was solicited to disprove the idea once and for all. As science historian Lisa Pollack has recounted, Wolfgang Wiltschko was asked to re-create an experiment conducted by a fellow student, who had put birds in a closed room without sunlight or stars but, to his surprise, discovered they could still orient. There were two possible explanations for this behavior: the birds used magnetism, or they used radio signals emitted from stars. Wiltschko thought the star hypothesis was the likely answer. He put European robins in a steel chamber that weakened the earth's geomagnetic field, and he kept them in it for several days to try and manipulate their internal clock. But when he tested them in an orientation cage, they were still perfectly oriented. If he reversed magnetic north, the birds could sense the change and switched the direction in which they tried to fly. Working with his wife and fellow scientist Roswitha, Wiltschko became convinced that birds used an inclination compass—the angle between the magnetic field and the horizontal plane of the earth—to navigate, and he conducted dozens of experiments with birds to prove it. Meanwhile, other studies emerged showing that sharks, skates, cave salamanders, snails, rays, and even honeybees seemed to possess a magnetic sense. By the early 2000s, scientists had shown that seventeen other species of migratory birds, as well as homing pigeons, used magnetic compasses.

The notion that animals have a bio-compass that can "read" the earth's geomagnetic field has now emerged as the most promising explanation of animal navigation. In addition to those marathon migratory species, nearly every animal that has been tested thus far demonstrates a capacity to orient to the geomagnetic field. Carp floating in tubs at fish markets in Prague spontaneously align themselves in a north-south axis. So do newts at rest, and dogs when they crouch to relieve themselves. Horses, cattle, and deer orient their bodies north-south while grazing, but not if they are under power lines, which disrupt the magnetic field. Red foxes almost always pounce on mice from the northeast. These organisms must all have some kind of organelle that functions as a magneto-receptor, the same way an ear receives sound and an eye receives space.

As evidence of such a capacity across species grew in the twentieth century, the allure of a universal theory for animal wayfinding based on magnetism increased. Could a biological compass based on magnetism explain the capacity for species like humpback whales to find their way?

Maybe. The only problem is that no one can find it.

———

The search for a bio-compass has now spanned nearly half a century, attracting biologists, chemists, and even physicists. But the anatomical structure, mechanism, location, and neural connections of animals' magneto-receptors remain a mystery. Kenneth Lohmann, an expert in sea turtle navigation, has called the search "maddeningly difficult." Magnetic fields, Lohmann wrote in *Nature*, pass freely through biological tissue, which means that the magneto-receptors could be located almost anywhere inside an animal's body. They may be so tiny as to be submicroscopic and dispersed throughout a body; it's even possible that magneto-reception could in fact be a chemical response, which means there's no single organ or structure devoted to it. "We are still crying out for *how* do they do this," the geologist Joe Kirschvink told me. "The compass is a needle in the haystack."

I met Kirschvink at a conference hosted by the Royal Institute of Navigation, which focuses on modern navigation by air, sea, river, and space. Every three years, their conference dedicated to animals brings together the world's foremost scientists to present their research. The year I attended it was held at Royal Holloway College, an ornate Victorian-era building a few miles southwest of London whose lavish interiors have appeared as sets in *Downton Abbey*. During tea and sandwich breaks it was clear that the conference was a reunion for researchers who have worked in the field for decades, and the mood was friendly. But there was a schism between the scientists.

In one camp were those who believed the bio-compass could be explained by magnetite, iron crystals in animal cells that enable the development of organs capable of detecting the geomagnetic field. In the second camp were those who believed magneto-reception is best explained as a biochemical reaction that is influenced by the geomagnetic field—a model of navigation that depends on quantum physics. Although some of these scientists maintained that the bio-compass might involve a combination of mechanisms, many scientists at the conference had focused entire laboratory budgets on proving one hypothesis over the other, turning the search for the bio-compass into a scientific race. There was a lot at stake, not only the distinction of solving a problem that science has thus far been unable to solve but also potentially widespread applications for technology and medicine. The discovery of the bio-compass mechanism, for instance, could give momentum to the emerging field of quantum biology, the idea that quantum mechanics is more than "the deep substrate on which biology exists" but is actually the mechanism behind many biological phenomena, or it could launch a new era of magnetogenetics—the ability to control molecules in cells using magnetic fields.

It was Kirschvink who discovered the mineral called magnetite, a naturally occurring oxide of iron, in honeybees and homing pigeons as a PhD student at Princeton University, and soon thereafter he presented magnetite as the basis of a bio-compass in animals. Just a few crystals of magnetite, he wrote, needed to be present in a cell in order to detect the geomagnetic field. Energetic and outspoken, he remains a firm believer in the magnetite hypothesis of animal navigation. It is, he explained, the most rational evolutionary pathway for migratory behavior seen in the full range of animals. Natural selection took something that worked marginally—the receptivity of magnetite to the magnetic field—and through mutation and gene replication made it better and better until it produced navigational wizards such as the bar-tailed godwit. "You have to be able to build these things step by step. You have to have something to select for it," he said. "You can do that with a magnet easily."

The presence of magnetite in so many animals does seem like sure proof of the bio-compass. Throughout the 2000s a lot of researchers honed in on the presence of magnetite in olfactory cells of rainbow trout and brains of mole rats, as well as in the upper beaks of homing pigeons, as potential magneto-receptors. But then a team of scientists at the Institute of Molecular Pathology in Vienna took a closer look, slicing the beaks of hundreds of pigeons and staining them to detect iron-rich cells, and they found big discrepancies in the number of cells present. Some pigeons had a couple hundred while others had tens of thousands. The likely explanation was that the cells were simply the product of an immune response in the birds' white blood cells. This doesn't mean that the magnetite hypothesis is dead—far from it. "One equivalent of a magnetic bacteria can give a whale a compass. One cell," said Kirschvink. "Good luck finding it."

Around the same time that Kirschvink was finding magnetite

in honeybees, a German physicist by the name of Klaus Schulten was looking at how radical pairs, two molecules each with an unpaired electron, could be responsive to magnetic fields. When the two electrons in a radical pair are correlated—in a state of either entanglement or coherence, meaning particles or waves affect one another even when separated over distances or split— the magnetic field is capable of modulating the electrons' spinning motion. Two years later, Schulten published a second paper suggesting that this phenomenon might be the basis of a biomagnetic sensor in birds, a kind of "chemical compass," he wrote, that was triggered when light caused an electron transfer reaction, generating radical pairs, that was then influenced by the external magnetic field.

For the next twenty years, no one knew where such a radical pair reaction might take place in animals. "It was clear that the radical pair mechanism was genuine," Peter Hore, a professor of physical and theoretical chemistry at Oxford University, told me, "but highly speculative that it might happen inside a bird's body." Then in 2000 Schulten proposed a newly discovered protein called cryptochrome, which was found in plants and believed to regulate growth during photosynthesis. Cryptochrome proteins are a kind of flavoprotein that is receptive to blue light; they have since been found in bacteria, the retinas of monarch butterflies, fruit flies, frogs, birds, and even humans. And they are the only candidate thus far that has the right properties for what some are calling the quantum compass.

Hore's research is focused on the behavior of radical pairs, and he became interested in testing the cryptochrome hypothesis after Schulten presented it. Genteel, with white hair and thin wire-framed glasses, Hore explained to me during the conference how difficult it is to design a "killer experiment." While researchers can show that radical pairs produced in the proteins are sensitive to magnetic fields, the weakest magnetic field capable of affecting cryptochrome is still twenty times *stronger* than the earth's magnetic field: no one has yet shown how these radical pairs can respond to earth's extremely weak geomagnetic field. At least part of the problem is that it has been nearly impossible to reproduce the actual conditions of the cell for experimentation. Hore predicts that proving cryptochrome as the bio-compass will require at least another five and perhaps as many as twenty years of research. If and when that day comes, it will constitute a critical contribution to a new field of quantum biology, the study of quantum effects in living organisms.

The idea that nature may have harnessed quantum mechanics in the course of evolution is compelling and controversial. There is now, for instance, evidence that quantum dynamics are involved in photosynthesis, when photons are absorbed and transferred to a cell's reaction center, creating electron excitation. Discovering more examples may lead to new quantum technologies. "The hope is that if these things are genuinely quantum biology," Hore said, "maybe we can learn lessons that would enable us to make better magnetic senses or more efficient solar cells using principles from nature."

———

The 2011 study of humpback whales concluded that magnetism alone couldn't explain the animals' migration, because there was a lack of any consistent relationship between their direction of travel and the magnetic inclination or declination. And I found a small group of researchers at the London conference who were skeptical of magnetism as a universal explanation for animal navigation. Kira Delmore is a young Canadian biologist at the Max Planck Institute for Evolutionary Biology, whose research focuses on the Swainson thrush, of which there are two different subspecies, each with a different migratory route. One travels along the west coast of North America to Central America and the other over the Midwest. Delmore wanted to know if by using a combination of geolocation gadgets and genetic sequencing she could find out whether the birds' different migratory behaviors were associated with particular genetic traits. In other words, could migratory orientation be explained by genetics? After years of research, her data showed that a bird's decision to go south or southeast *does* have a genetic basis. "Migration is such a complex behavior, the idea that there is a gene that might say whether I should go left or I should go right, if that's genomic, it boggles my mind," she said.

Hugh Dingle has proposed that evolution created what he calls "migratory syndromes," made up of a combination of behaviors and physiologies such as the suppression of maintenance activities, the use of fat as fuel, and navigation, and that each syndrome is shared across species and defines them as migratory. The definition of the word *syndrome* is revealing. It comes from the Greek *sun*, meaning "together," and *dramein*, "to run." But in the mid-sixteenth century, *syndrome* also became a medical term referring to a condition, illness, disorder, or sickness. Maybe that is what migratory syndrome feels like for the snow goose driven north to the Arctic each spring: not like a choice but like a powerful compulsion. In 1702 the ornithologist Ferdinand von Pernau described feathered migrants as forced to depart because of a "hidden drive at the right time." In the late eighteenth century, naturalist Johann Andreas Naumann kept golden orioles and pied flycatchers in a room, cutting a small hole in the door so that come winter he could watch them become agitated and restlessly try to escape their captivity. Charles Darwin once wrote about how John James Audubon kept a "pinioned wild goose in confinement, and when the period of migration arrived, it became extremely restless, like all other migratory birds under similar circumstances; and at last it escaped." What inner longings steer a goose on its long journey north? What forces propel it there? These feelings or intuitions aren't alien to our own human experience. Indeed the Swiss-American psychologist Elisabeth Kübler-Ross likened the hidden drives in animals and men to one another. As a young woman she traveled by ship from Europe to

America, a place she had never been before, and during the journey wrote in her journal, "How do these geese know when to fly to the sun? Who tells them the seasons? How do we, humans, know when it is time to move on? As with the migrant birds, so surely with us, there is a voice within, if only we would listen to it, that tells us so certainly when to go forth into the unknown."

NAVIGATION MADE US HUMAN

In a modern apartment on the north side of Iqaluit, I stood in front of a large contour map scattered with traditional Inuktitut place-names, chewing on caribou jerky whose center was still raw. Next to me was Daniel Taukie, a thirty-seven-year-old hunter from Cape Dorset, a community renowned for its arts on the far western coast of Baffin Island. Taukie pointed to a lake that we were going to travel to that day, about eighteen miles north of where we were standing. In English it is called Crazy Lake; in Inuktitut its name is Tasiluk Lake, meaning odd-shaped and shallow, a lake pretending to be a lake. Taukie was hunting ptarmigan, a white-feathered bird whose smooth, dark-violet flesh is best eaten raw. I was in pursuit of *inuksuit* (the plural of *inuksuk*), stones of various sizes and shapes that have been placed upon each other to create a structure. In Inuktitut the word *inuksuit* means "that which acts in the capacity of a human." Some are made up of dozens of stones, some have just two or three, and their purposes are diverse—navigation, hunting, or caching meat. It was Solomon Awa who had told me I could find some very old *inuksuit* near Crazy Lake, and Taukie had volunteered to take me.

I had met Taukie at Nunavut Arctic College, where he had recently graduated from a two-year training course in wildlife management. Handsome and affable, Taukie is an ardent hunter and an indefatigable source of knowledge about traveling on the land. "That's what he loves, that's what he does," said Jason Carpenter, one of his teachers. "Vacation in Florida? No, he wants to go to this floe edge, he wants to get a seal pup." Taukie's particular passion was wolves, but he hunts caribou, walrus, polar bear, and fox. In 2009 he had participated in a bowhead whale hunt in Cape Dorset, the first in the community in a hundred years. The hunters had spotted the whale twenty-five miles offshore, and as the lead harpooner, Taukie had taken the first shot at the fifty-foot whale. Once the whale was dead, it took nearly eight hours to tow it to shore, where some five hundred people gathered to greet the hunters and harvest the animal.

"For your conscience, I won't bring a GPS," Taukie said with a smile. He considered the device a bad way to travel anyway. "You can't take shortcuts, there's just one route. Sometimes it doesn't know a cliff, so I can't depend on it too much. When I do, it's on a flat plain or in a whiteout." As a kid hunting caribou with his father and other community members, it was typical for Taukie to travel four or five hours through the night without any maps or GPS to reach the herds. Today the widespread use of fast snowmobiles coupled with GPS has increased the hunters' reach, and caribou numbers have crashed in recent years. "Snowmobiles can travel farther than we used to go on a daily basis. We were catching more and more out in places we never used to catch them," he told me. "I can honestly say I was part of the decline. I was getting five caribou every two weeks."

Listening to Taukie's stories about his travels, I had the impression that much of Baffin Island was his backyard. He told me about a solo journey he once made from Iqaluit to Cape Dorset, a nineteen-and-a-half-hour trip by snowmobile, without sleep, following old trails, some of which he had only heard about. Looking at the map on the wall was for my benefit. We went outside and packed his wooden *qamutiik*. The sun had been up since three a.m., and the blue sky was so sharp and the air so cold that the land around us seemed to crackle and spark. We put backpacks, a cooler, a shovel, thermoses of hot water, and some bannock under a blue tarp and carefully tucked its edges as though making a bed. We threaded rope through the gaps in the sled's runners and around the entire bundle and cinched it tight. Taukie wore polarized sunglasses, the shape of which was outlined on his tanned face. He secured an extra five-gallon container of fuel on the sled and slung two rifles around his chest. Then we jumped on the snowmobile and pulled out from behind his apartment building, heading up into the hills and valleys that border Iqaluit to the east, an infinite terrain of rock and snow cast in blinding sunlight, to hunt.

When did humans lose the biological hardware that allows so many animals to navigate with such precision? Did the hippocampus replace it? As neuroscientist Howard Eichenbaum pointed out to me, there's no fossil record of the hippocampus. We don't know what it did a hundred thousand years ago. Scientists can only guess at its evolution. But the fact that it is very, very old, at least hundreds of millions of years, is a significant clue. Even birds, which last shared an ancestor with humans 250 million years ago, as well as amphibians, lungfish, and reptiles, have what is called a medial pallium. Similar to the mammalian hippocampal formation in vertebrates, the medial pallium is also involved in spatial tasks in these species, raising the possibility that certain properties of spatial cognition were conserved as organisms diversified and split, while other properties adapted to particular ecologies or selective forces. But despite the profound evolutionary commonalities between humans and other vertebrates and the way the hippocampus relates to cognitive functions of memory and navigation, the question remains: why did we make such a leap in terms of hippocampi's size and role in our lives? Or as psychologist Daniel Casasanto puts it, "How did foragers become physicists in the eye blink of evolutionary time?" Perhaps it was exactly what Taukie and I were heading out to do—hunt—that contributed to our unique navigation strategies and intelligence, eventually engendering a practice we consider the most human of all: storytelling.

———

Sherlock Holmes and Sigmund Freud would seem to have little in common. One is a fictional crime detective, the other a physician who founded the field of psychoanalysis. And yet according to the Italian historian Carlo Ginzburg, Holmes and Freud were remarkably similar in that their work involved the mastery of a particular type of information—what is called conjectural or evidential knowledge. Ginzburg describes this kind of knowledge as the "ability to construct from apparently insignificant data a complex reality that can not be experienced directly." This data is most often made up of traces from the past; a footprint, an artwork, a fragment of text. In Freud's work the traces were symptoms observed in his patients; in Holmes's case the traces were clues gathered from crime scenes.

In the late nineteenth century, Ginzburg argues, evidential knowledge emerged as an epistemological paradigm that influenced a broad scope of disciplines from art history to medicine to archaeology—and figures like Arthur Conan Doyle and Freud. Despite its influence during this period, Ginzburg believed that its roots were much much older; he argued that it originated in our species' skill as hunters. In his 1989 book *Clues, Myths, and the Historical Method,* Ginzburg writes,

> Man has been a hunter for thousands of years. In the course of countless chases he learned to reconstruct the shapes and movements of his invisible prey from tracks on the ground, broken branches, excrement, tufts of hair, entangled feathers, stagnating odors. He learned to sniff out, record, interpret, and classify such infinitesimal traces as trails of spittle. He learned how to execute complex mental

operations with lightning speed, in the depth of a forest or in a prairie with its hidden dangers.

For Ginzburg, the hunter, detective, historian, and physician are all part of a sign-reading paradigm. In the hunter's case, by reading and deciphering tracks, the hunter is able to produce a narrative sequence, and Ginzburg posits that the idea of narration might have originated in a hunting society. "The hunter would have been the first 'to tell a story' because he alone was able to read, in the silent, early imperceptible tracks left by his prey, a coherent sequence of events."

Some linguists argue that the first human protolanguage that eventually led to symbolic language might have come from efforts by foragers and hunters to reconstruct a situation or scene for somebody else—the location of an animal or water source, say. In his book *More Than Nature Needs*, linguist Derek Bickerton described how language could have evolved as a solution to displacement, an ability to describe things that are not physically present and also to coordinate a plan of action. Bickerton calls this particular scenario "confrontational scavenging"—recruiting a big enough group of individuals to travel to the site of an animal carcass and drive off animal competitors to get meat. Presumably, spatial descriptions and directional terms were critically important for confrontational scavenging, and so these primordial conversations might have contained a lot of navigational information and created the need for an ever-expanding navigational vocabulary. According to this theory, the birth of language was just the human version of a bee dance. "[T]he notion that the world might consist of nameable objects was literally inconceivable to animal minds," says Bickerton. "Recruitment for confrontational scavenging forced prehuman minds to accept the notion. When a prehuman who had just discovered a carcass appeared before conspecifics gesturing and making strange sounds and motions, the motive could only be informational: 'There's a dead mammoth just over the hill, come help get it!'"

Scavenging led to chasing animals down and, eventually, trapping them, and traps require a great deal of abstract and complex thought. The British social anthropologist Alfred Gell believed that while chasing animals physically pitted the hunter and animal against one another as equals, the trap created a hierarchy of the hunter over the victim. Traps are designed to take advantage of the animal's behaviors, they create a "lethal parody of the animal's *Umwelt*," Gell said. A trap is a model of its creator and a model of its prey, embodying a "scenario, which is the dramatic nexus that binds these two protagonists together, and which aligns them in time and space." At the cognitive level, traps require a high level of evidential knowledge: understanding the movements, habits, and lives of animals in order to make predictions of their future movements. And traps are powerful signs: "We read in it the mind of its author and the fate of its victim."

In the past everyone probably needed some tracking skills for survival; today it is a lost skill practiced by a minority. But you don't have to have trapped or hunted wolves to realize how profound and ubiquitous the skills of "tracking" are in our everyday lives. Our existence depends on thousands of instances of inference and deduction that allow us to draw conclusions about other people and things, what has happened and what will happen, about causes and consequences. We continually tell ourselves stories and test them against reality. The evidential paradigm described by Ginzburg seems to be at the root of so much human thought.

I spoke to an evolutionary philosopher, Kim Shaw-Williams, who believes that navigation might have been the original purpose of what he calls "trackways reading" and its role in hunting, tracking, and trapping. Shaw-Williams grew up in the backwoods of northwestern Canada, and he has had an eclectic career, one that has included an ecology degree, possum hunting in the New Zealand wilderness, and working on film sets before enrolling in a PhD program at the University of Victoria in Wellington. It was during this period that his experiences as a young kid trapping animals took on new significance. He recalled an epiphany he had one morning before catching the schoolbus, doing a routine check of trap lines near his house. "There was a light bit of snow the night before, and what happens with rivers in winter is the winds blow up and down, creating a hard crust on the snow, and the animals use them as a highway," he told me. "I was wandering along the river and saw these tracks, coyotes, foxes, and all these different animals. And I had this eureka moment that before we became human, we were *reading* these things." With this thought came the feeling that he was accessing a conduit in consciousness to our ancient ancestors. "It was a real buzz," he recalled.

From his home in New Zealand, Shaw-Williams described to me how he now thinks that the ability to read animal tracks was a selective trigger 3.5 to 3 million years ago, spurring the cognitive evolution of our earliest widely accepted hominin ancestors, leading to changes in our lineage's genetics, morphology, cognition, and behavior. Subsequently, some 2.3 million years ago during the early Pliocene, the human brain started to undergo encephalization, an increase in its relative size, which Shaw-Williams believes was required for storing memories in order to effectively forage. "What the first encephalization was about was socio-ecological information. The memory for trails, memory for techniques, and being able to navigate between one place and another," he said. Tracking footprints—of animals or people—helped navigation because this sign essentially constituted trails that could be followed to get from one place to the next, find someone who was lost, or return to the group. These practices brought about dramatic and complex cognitive shifts. "To track, you have to have mental representations of other agents moving through an environment," he said. "Once we needed to become more efficient at foraging journeys, navigation became a bigger thing. Orienteering, taking an egocentric or allocentric point of view, triangulating them, and keeping track of your journey."

Shaw-Williams calls his ideas about evolution and tracking the "social trackways theory." It suggests that hominids are animals that learned to "read" the tracks of other hominids and animals and eventually infer meaning about events that happened in the past from these symbols. This enabled them to predict future behaviors based on these stories and use them to find one another, avoid predators, and successfully hunt prey. The ability to read tracks eventually led our ancestors to create first their own artificial signs and symbols to mark trails, then signed and spoken languages, and eventually the written word. (What is a book but the trail of words on paper left behind by a wandering mind?) The social trackways theory posits that humans are the only species to have undertaken the visual analysis of patterns of indentations left behind by other people and moving animals, a skill that created a unique cognitive niche: the narrative mind. A tracker imagines being "in the mind and body" of the author of the trackway, and then creates a narrative. To do this, they would need an egocentric self-referentiality and an allocentric sense of another's point of view. "I have to mentally picture the target animal approaching the whole trap site in the future, essentially by imagining myself to be in the body and mind of the target animal," said Shaw-Williams. "When setting traps or snares everything is done according to the perceptions and possible mental state of the animal I am imagining I am being while I am setting the trap." Within these strategies are the origins of characteristics that came to define our species as unique from others: self-projection, role-playing, sophisticated tool use, future planning, and symbolic communication.

This sequence of cognitive development also describes the birth of autonoetic consciousness, the capacity to be aware of one's own existence as an entity in time. The word *autonoetic* comes from the ancient Greek word for "perceptive." The term is sometimes used in the context of schizophrenia studies; patients who believe that their own thoughts are actually coming from an outside source have autonoetic agnosia. In the 1970s, the influential experimental psychologist Endel Tulving brought attention to autonoetic consciousness when he distinguished

episodic memory, the recollection of events of the past, as separate from semantic memory, the conscious access to context-
dissociated facts. To Tulving, episodic memory is the glue in a
system that allows humans to maintain a coherent sense of
identity and self through the conjunction of subjectivity, autonoetic consciousness, and experiences. This episodic memory system allows us to locate ourselves in time, to travel back
to the past and forward into the future (what's known as pros-
copy or prospection).

Tulving believes these abilities were what separated humans
from animals. "Although common sense endows many animals
with the ability to remember their past experiences," he has written, "as yet there is no evidence that humanlike episodic memory—
defined in terms of subjective time, self, and autonoetic
awareness—is present in any other species." It's not that animals
can't remember the what, where, and when of particular events—
the studies of scrub jays discussed in the previous chapter show
that they can—it is that these are more likely simple episodic-like
abilities, different from the profound autonoetic consciousness
that is a hallmark of human cognition. Some other researchers see
in the results of recent rat maze studies enough evidence to question this argument. New Zealand psychologist Michael Corballis
now thinks that hippocampal activity in sleeping rats, in which
they seem to be replaying but also anticipating future excursions,
may prove that mental time travel is present in other animals,
and it is only the degree of complexity in humans that is unique.
But Tulving argues,

> Lest someone worry about the current political correctness
> of such an assertion, let me hasten to remind such a person
> that many behavioral and cognitive capabilities of many
> non-human species are equally unique to those species:
> echolocation in bats, electrical sensing in fish and genetically
> determined navigational capabilities of migratory birds
> are examples that come quickly to mind, but there are

many, many others. Indeed, it is these kinds of abilities—unfathomable by common sense, but very real in fact—that allow one to remain a sceptic about episodic memory in birds and animals: evolution is an exceedingly clever tinkerer who can make its creatures perform spectacular feats without necessarily endowing them with sophisticated powers of conscious awareness.

According to the trackways reading hypothesis, once tracks became a much-used strategy for navigation, foraging, finding water, remembering routes, and hunting for animals, it led to humans creating rich mental maps of territories and routes based on narrative memory of previous experiences and the experiences of others. Our memory capacities grew, and we amassed more natural history information—the changing seasons, migration patterns for animals, breeding cycles, habitats. Learning all of this took time and energy, potentially extending childhood and juvenility and their periods of neural development. Out of this process emerged a creature that could begin to organize its experiences in space and time, to navigate farther, to build complex maps and sequences in the brain, and, eventually, once they harnessed symbolic communication and then language, to communicate these geographic and biographical narratives to others.

———

Shortly into my hunting trip with Taukie I began to encounter a common problem for people unused to traveling on the Arctic landscape: my eyes seemed to lose the ability to perceive scale and therefore distance. Living in the eastern United States, I was accustomed to my visual field being filled with tall objects—trees, buildings, lampposts—which my brain unconsciously interpreted as reference points to tell how much space existed between me and places on the landscape. Up in the Arctic the tallest trees were twelve-inch willows, more like bonsai trees than the oaks and pines I was used to. Hills and mountains can be reference points, but once we entered the tundra, the biggest objects were rocks. At times I thought a rock was hundreds of yards in the distance but it turned out to be only a few yards away. The monotony of the landscape was scrambling my brain. Taukie didn't have this problem. "It may look all the same to you, but it's very detailed in our heads," he said. "When we're taking in details in our head, we try to look at the little stuff. When we travel past something we look behind us because it looks different from that angle. Every detail we are trying to put into our heads."

Even while driving at fifty miles an hour Taukie methodically scanned the landscape for any flashes of ptarmigan while simultaneously paying attention to looming rocks and crevices ahead that threatened to overturn the snowmobile. I tried to keep track of our route in my head, but I soon became disoriented. I was further distracted by the drop in air temperature as we climbed in elevation; my face began to freeze, and I pushed the front of my coat up over my nose. If I left my hands ungloved for longer than a minute, my fingers started to ache. After a while Taukie slowed the snowmobile and then stopped to point at the ground. "I use these to tell direction," he said. I looked closely and noticed a few rocks. "See the snow behind the rocks?" said Taukie. "It shows you where the dominant wind is coming from." Sure enough, on the lee side of the rocks was a ridge of snow. Broad at the base, it narrowed to a point, shaped by the blowing wind hitting the rocks and then wrapping itself around the obstruction. This was another kind of *sastrugi*, not *uqalurait*—the tongue-shaped formations found on sea ice—but *qimugiuk*, a tapered ridge found behind protrusions in the ground. In this area the dominant wind was coming from the northwest, and each of the ridges showed that it came from that direction. As we continued on the snowmobile I started to notice that *qimugiuk* were everywhere. Even a small rock that could fit in my palm had a small ridge pointing southeast. We passed a large hill to our right and I realized that it, too, had a giant *qimugiuk* behind it. A whole mountain, it seemed, could have a *qimugiuk* the size of a fallen skyscraper. I looked at the placement of the sun; this time of year it was rising in the south and setting late at night in the north. Between the *qimugiuk* and the sun, I suddenly realized that I now knew exactly where Iqaluit was and what direction we were going, no matter how many twists and turns we took on the snowmobile. In these near-perfect conditions, it seemed nearly impossible to get lost. For the first time, I felt as though I was reading the landscape.

Eventually we came up the steep side of a slope and rested on the top: we were on the lip of a gentle bowl, and below us was an enormous flat plain covered in snow. This was Crazy Lake, frozen solid. We followed along the rim until we both saw in the distance a distinct pile of stones—an *inuksuk*. Taukie drove the snowmobile up beside it, and we got off to look more closely. Just under three feet high, it was made of half a dozen cantaloupe-sized rocks upon which a single rectangular rock, perhaps two feet long, was placed. Its bottom side was nested into the smaller rocks, while its top was flat and gently sloped. On the very top, snugly fit so that it could not be moved without great force, was a thick pinkish-hued stone. Taukie took his hands and began to feel the flat rock, seeking its shape tactilely. He noted that it was an older *inuksuk* because of the patches of lichen growing over it, growth that takes decades in the Arctic climate. "You can tell the middle rock, the shape of it is pointing." Sure enough, I could see how the rock narrowed slightly on one end and pointed toward the southwest. This was vaguely in the direction of Iqaluit. The *inuksuk* likely predated the permanent settlement of the town, so maybe it was pointing to the direction of the coast; we couldn't know for sure.

Norman Hallendy, an Arctic researcher who spent decades studying *inuksuit,* describes them as stones that act as mnemonics for travelers. He documented eighteen kinds in southern Baffin Island alone, each with a different name and use. An *aluqarrik* marks a spot to kill caribou, and an *aulaqqut* is for frightening caribou. Some indicate the depth of snow, the location of a food cache, dangerous ice, a fish spawning spot, or sites for mining pyrites or soapstone. Some hunters built entire fences of *inuksuit* that would guide the caribou herds toward their spears. Other *inuksuit* were designed with hanging bones to catch the wind and send sound across the land to distant travelers.

One of the important purposes of *inuksuit* is to help with navigation. The rocks are compiled in order to indicate the best way to get home, or the direction of the mainland from an island, or a place that is below the horizon, or where seasonal animals are located. To this end, they can have very different designs. A *turaarut* points to something, while a *niungvaliruluk* has a window that a person can look through to align themselves with a distant goal. Another kind of *inuksuk* is called *nalunaikkutaq*, which translates into English as "deconfuser." Hallendy has written that for the Inuit, "the faculty of visualization—being able to record in the mind every detail of the landscape and the objects upon it—used to be essential to survival. . . . Part of this skill was the ability to memorize the location of places in relation to one another, and in stretches of featureless landscape, an inuksuk was a great helper. Skilled hunters memorized the shapes of all the inuksuit known to the elders, as well as their locations and the reasons they had been put there. Without these three essential pieces of information, the messages the inuksuit contained were incomplete."

For piles of rock, *inuksuit* can be remarkably durable. In northern Baffin Island near Pond Inlet, Carleton University archaeologist Sylvie LeBlanc has studied a chain of *inuksuit*—around one hundred stone objects marking a route from a lake to the mouth of the bay, likely used by people representing every Arctic culture from Pre-Dorset to Thule to Inuit. Measuring ten kilometers in length, it is thought to be the longest navigational system of its kind ever documented and at least forty-five hundred years old.

There is no Rosetta Stone for decoding *inuksuit*. Each one is unique, assembled by a person who studied the rocks available to them, examining their shape and weight, and then joined them together to create a message. To decipher them, you have to infer the intelligence at work, projecting yourself into the mind of the object's creator. Where were they going? What did they want? What were they trying to say through the stone? Taukie said he implicitly trusts the wisdom of the people who created them, even if it can take him some time to discern their intent. Many times he's been able to infer a better route by studying their placement. Others who travel on the land told me the sight of an *inuksuk* has brought a profound mental comfort and relief even if it is not used as a directional or hunting aid. The sight of a pile of rocks indicates someone has simply been here before even if it was hundreds of years ago. One Iqaluit resident told me they traveled sixteen days from Iqaluit to Pangnirtung by dogsled on a trail that included some four hundred *inuksuit*. Others described how in whiteout conditions, *inuksuit* had an uncanny tendency to appear just as they began to experience the dread that they had surely lost their way.

———

Back on the snowmobile I pointed out stones to Taukie. "*Inuksuik?*" I'd ask. "No, I think that's just some rocks," he'd say politely. It was trickier than I realized to distinguish the work of a human mind from the vagaries of nature: wind, snow, physics, chance, and time created their own quirky designs, a bigger stone precariously balanced atop a smaller one, for instance. Part of discriminating between what might be a simple *inuksuk* and nature's creations was trying to glean the riddle of human ingenuity—subtle, rational, peculiar—in the landscape.

Taukie hadn't seen any ptarmigan and was determined to shoot some, so we headed farther east, deeper into the interior and along the valley floors that run on a north-south axis into the bay, constantly scanning the hillsides for the flutter of white wings. Twice I called out "ptarmigan!" only to realize a second later that what I was seeing was the enormous wingspan of snowy owls drifting close to the ground as they, too, hunted. "That explains why we're not seeing any birds," Taukie said, disappointed. I was thrilled. We saw a snow bunting flit across a narrow trail in front of us, and I remembered that Harold Gatty had written about an explorer who marveled that his Greenlandic travel companions found their way home through fog by identifying the songs of the male snow bunting.

In the late afternoon we stopped to drink tea in a valley floor. The snow had begun to melt there, and the tundra was run through with rivulets of fresh water. We sat in a soft field of crowberry bushes and picked some of the small purple berries, plump with moisture from the thaw. Not far from us was an old tent circle, the discarded stones left by previous generations of Inuit travelers. We discussed Taukie's membership in a core group of hunters in Iqaluit, for whom hunting is not just a weekend pastime but a livelihood, practice, and daily interaction with Inuit tradition. But making a living as a hunter today is challenging. Harvested animals are traditionally viewed as gifts that offer themselves to the hunters based on their skill and integrity. Meat is the property of the community rather than the individual, and hunters often support a network of elders and families, providing fresh seal and fish to them for free. By sharing rather than selling their meat, hunters sometimes have to go on welfare or hunt less in order to work; the subsequent lack of traditional food in Nunavut has been cited as a contributing reason for the territory's persistent food insecurity.

"Right now hunters go out and bring back their harvest and just give it away. That's great except there is no way to make that an economically sustainable chain," said Will Hyndman, a consultant to Nunavut's Department of Economic Development and Transportation, in a later conversation. "In the past you'd share, but part of your harvest would be to feed your dogs and use those materials to go back hunting again. Now . . . you come back with three hundred pounds of fish, but you have no way of repairing your snowmobile. . . . Here all the best hunters have the shittiest gear. It's tough, you can't earn a livelihood as a hunter." There are some signs that the resistance to selling meat for cash has eased. For instance, there is a Facebook group called Iqaluit

Swap/Share with twenty thousand members, where freshly harvested meat is often posted for sale alongside handmade
children's clothes or used appliances.

Taukie and I smoked cigarettes and then packed up our thermoses. After a few more hours of traveling toward the coast, we
decided to forgo the hunt when we came upon a high hill crossed
with snowmobile tracks. Taukie gunned the snowmobile up its
steep face then turned it sharply around at the top. For a moment
the snowmobile lost contact with the ground and the g-force
made our stomachs lurch. We landed with a WOMPF, laughing
on the way back to the bottom before catching our breath and
doing it all over again. At the end of a valley where the land met
the sea, we tried to find a way down the steep cliffs and onto the
ice before retreating back through the tundra. I saw an Arctic
fox in the distance, traipsing in the shadow of some rocks, pausing periodically to stop and look at us looking at her. When we
finally pulled up behind the apartment we were hungry. Taukie
put a piece of cardboard on the kitchen floor and began slicing
cubes of raw caribou fat and strips of red steak from a previous
hunt. The fat was like butter delicately flavored with musk. Wind-
burned and exhausted from the cold, I went home to sleep through
the lingering twilight that is night in the Arctic spring.

———

The next day I walked half a mile to downtown Iqaluit to visit the
offices of the Inuit Heritage Trust, an organization launched in
1994 following Nunavut's independence as a territory with the
mission of protecting Inuit identity through the preservation of
archaeological sites and ethnological research. One of its man-
dates is to record the traditional Inuktitut place-names so that
they can undergo an official process and be included on all future
maps generated by the Canadian government. Inside the offices I
sat down with Lynn Peplinski, the manager of the place-names
project. Tall and enthusiastic, Peplinski is a veteran dogsledder,
and her work requires her to visit far-flung communities across

the territory and spend hundreds of hours poring over maps with elders and hunters. So far the Trust has produced ten thousand traditional place-names, and they haven't finished yet.

Place-names are an intrinsic aspect of Inuit wayfinding: learning them helps travelers know where they are and remember the sequence of routes on land and along coastlines. And the Inuit were prolific namers who focused on the fine details of topography: inlets, hills, rivers, streams, valleys, cliffs, campsites, and lakes. In contrast, Peplinski explained to me that Europeans like Frobisher focused on naming entire landmasses for the purpose of mapping and conquest, often after individuals. "For the explorer, they want to map the area, and they need to name the landmass that blocks the waterway," she said. "For the Inuit who want to use the land, [that way] doesn't make sense."

Some of these descriptors were very specific. An island might be called "shaped like an animal's heart," while another would be called "point of a parka hood." The pictorial descriptions helped people recognize places they might never have been to before. "If you are an Inuk who understands the language," said Peplinski, "chances are you'll be able to have a mental image of what that place looks like from the name." In testimony from the three-volume report prepared in the 1970s that became the basis of Nunavut's land claim, one woman explained how place-names work: "Sometimes we name them on account of their size or because of their shape. The names of places, of camps and of lakes are all important to us, for that is the way we travel—with names," said Dominique Tungilik. "We could go anywhere, even to a strange place, simply because places are named. . . . Most of the names you come across when you are traveling are very old. Our ancestors named them because that is where they travelled."

Peplinski and her team only record names that are still "alive," meaning they come from the firsthand knowledge of elders who inherited them orally. But future generations of Inuit may very well inherit many of these place-names from the official government maps or Google. Right now, searching Baffin Island on

Google Maps reveals a view of the territory not very different from the colonizing Europeans: it's mostly blank. Of course the coastlines are represented in minute detail thanks to the exacting accuracy of satellite imagery, but the land itself appears to be barren except for tiny islands of habitation in towns like Iqaluit. The place-names project will change that. Once the territory's legislature stamps the names compiled by Peplinski and her team, Google will be legally obligated to include them. Baffin Island and the rest of the eastern Canadian Arctic will be seen for what it is: a landscape traversed for thousands of years by people.

There is both a symbolic and practical importance to making these names official in the twenty-first century: it not only recognizes the fundamental legitimacy of Inuit knowledge and tradition, but it also guarantees that even those young people equipped with smartphones and gadgets can still learn the names their ancestors used to navigate. Ideally they might learn in the same way that Awa and Taukie did: on the land, directly through experience. Awa told me that learning from maps and in the classroom can only go so far in sustaining wayfinding practices. Getting out on the land, cultivating an independent relationship to it and a memory of its geography, is how traditions will stay alive. Awa thinks that the shift in education practices (more than snowmobiles or GPS arguably) has disrupted wayfinding skills in the younger generation. "People who are not taught the right way tend to get lost," Awa said. "If you want to learn how to navigate, go out there."

The resilience and fate of Inuit traditions is a topic that came up repeatedly during my time in Iqaluit. Like so many Native American and First Nation communities, the Inuit have higher rates of alcoholism, diabetes, depression, and domestic violence than other populations do. The land claim agreement enshrined the Inuit right to self-determination in law, but the consequences of colonialism, forced relocation, and residential schools, which divorced a generation of children from their culture, are still felt.

Many people told me that the traditional wayfinding skills

that hunters and the Inuit had mastered so completely would soon be gone, if they weren't already functionally extinct, the victims of permanent settlements and technology. When I arrived in town, everyone was still talking about the most recent search-and-rescue operation for three hunters who were found one hundred miles south of Iqaluit. This was puzzling because they had been originally traveling to Pangnirtung on a route that goes north. The hunters' explanation was that they had become disoriented by bad weather and bunkered down in an abandoned cabin. Then they followed their smartphones' GPS for *two days* in the wrong direction before realizing that it wasn't working properly, a shocking mistake that might have cost them their lives. The rescue operation cost over $340,000 CAN.

Yet my experiences in Nunavut led me to doubt the many dire predictions I heard for Inuit cultural survival. Traditions have been modified, transformed, or adapted but seemed far from forgotten. One night I went to the Storehouse Bar in Iqaluit, a popular hangout with flatscreen TVs and pool tables. It was impressively packed for a Wednesday; women and men of all ages danced under a splay of red and blue lights to electronic music, and the line to buy drinks was so long that everyone bought them two at a time. I ordered Molsons and whiskey and sat at a table with Sean Noble-Nowdluk and his friend Tony. They had been best friends since they were little kids and went to the same schools. Now they hunted together; both had the tell-tale sign of many hours spent under the Arctic sun—dark tan lines around their eyes delineating the shape of their sunglasses.

Spring hunting was in full swing, they told me, and they'd seen their first goose. The sea ice was so smooth they rode across the bay on their snowmobiles at seventy miles per hour, practically flying. It was Sean's twenty-first birthday; it had been a good day. Hunting was clearly the center of their lives. Both spoke with pride and reverence of learning to hunt from their relatives as children and talked about all the places and animals they'd seen. We pored over pictures on their smartphones. "Here's my little brother with his first polar bear," said Sean, pointing to a photo of a magnificent white bear with an expertly aimed, single bullet hole through its lungs. Sean's brother was nine when he killed it. They showed me a picture of a baby seal on the ice looking at the camera with large, liquid black eyes. I felt a sudden pang of regret that it had been killed, which I admitted. "We're taught never to disrespect animals. Ever," Sean said.

We left our drinks for a moment to go outside in the cold, where young and old stood around smoking together, many in T-shirts despite the below-freezing air. "How do you guys find your way out there? Do you use GPS?" I asked. "No!" Sean said, surprised at the question. "We know our way." He talked about using landmarks and *sastrugi*, of learning the routes from his dad. We went back inside, where they began to wax about the epic hunting feats of their friends in communities farther north like Arctic Bay and Grise Fiord, of a friend who sometimes drove his truck on sea ice and had once caught a beluga whale by the tail with his bare hands. The stories were full of youthful swagger and glory. When I left to walk home around midnight they were still drinking and celebrating. But the next morning, they were out hunting again, Sean with his new birthday gift: an anniversary-edition Ruger .22 to replace the rifle he'd hunted with since he was five years old. Back south in the United States I continued to follow their spring hunting travels on Facebook. "Done so much hunting in the last couple of months," wrote Sean one day, "that I really do feel I'm home out there rather than in town."

THE STORYTELLING COMPUTER

Our brains seem designed to conjure narratives from experience. And for eons, different cultures have used narrative sequence and story to transmit navigational knowledge. The anthropologist Michelle Scalise Sugiyama spent a year researching oral traditions of foraging societies and discovered how widespread this phenomenon is across the world. In her analysis of nearly three thousand stories from Africa, Australia, Asia, North America, and South America, she found that 86 percent contain topographic information—travel routes, landmarks, the locations of resources like water, game, plants, and campsites. She argues that the human mind—initially designed to encode space—found a way to transmit topographical information verbally by transforming it into social information in the form of stories. "Narrative serves as a vehicle for storing and transmitting information critical to survival and reproduction," she reported. "The creation of landmarks by a human agent or agents in the course of a journey is a common motif in forager oral traditions. . . . By linking discrete landmarks, these tales in effect chart the region in which they take place, forming story maps."

Examples of practices for creating story maps abound in Native American cultures from the Mojave to the Gitksan. In 1898

Franz Boas (who conducted his first anthropological fieldwork on Baffin Island) described a common feature of the stories of the Salish peoples, a character known as a "culture hero," "transformer," or "trickster" who gives the universe its shape through his travels and whose adventures are passed down through generations. A few years earlier, Richard Irving Dodge, an American colonel who spent thirty-four years in the American West, described an amazing capacity for memorizing orally transmitted topographical information among the Comanche in his 1883 book called *Our Wild Indians*. As a young boy, Dodge's guide had been kidnapped and raised by Comanches, and he told Dodge that a few days before young men set out on a raid, an older man with knowledge of the country would assemble them for instruction. As Dodge recounts,

> All being seated in a circle, a bundle of sticks was produced, marked with notches to represent the days. Commencing with the stick with one notch, an old man drew on the ground with his finger, a rude map illustrating the journey of the first day. The rivers, streams, hills, valleys, ravines, hidden water holes, were all indicated with reference to prominent and carefully described landmarks. When this was thoroughly understood, the stick representing the next day's march was illustrated in the same way, and so on to the end. He further stated that he had known one party of young men and boys, the eldest not over nineteen, none of whom had ever been in Mexico, to start from the main camp on Brady's Creek in Texas, and make a raid into Mexico as far as the city of Monterey, solely by memory of information represented and fixed in their minds by these sticks.

In her description of Pawnee migrations across the Midwest plains, anthropologist Gene Weltfish describes how each band followed a preferred route, some of which had few identifieable landmarks and were easy to get lost in. To successfully navigate, "the Pawnees had a detailed knowledge of every aspect of the land," she wrote in *The Lost Universe*. "Its topography was in their minds like a series of vivid pictorial images, each a configuration where this or that event had happened in the past to make it memorable. This was especially true of the old men who had the richest store of knowledge in this respect." Similarly, the cultural and linguistic anthropologist Keith Basso wrote in his book *Wisdom Sits in Places* that the Apache frequently cite the names of places in sequence, re-creating a journey. One day Basso was stringing barbed wire with two Apache cowboys when he heard one of them talking to himself quietly, reciting a list of place-names for nearly ten minutes straight. The cowboy told Basso that he "talked names" all the time, that it allowed him to "ride that way in my mind." Basso, who spent decades studying the place-naming practices of the Western Apache people, called place-names "intricate little creations." They are small words that do large tasks, one of which is to aid in navigation. For example, an Apache name like Tséé Biká Tú Yaahilíné translates to "Water Flows Down a Succession of Flat Rocks," a literal description of the place itself.

In the Apache settlement of Cibecue, Basso mapped forty-five square miles and recorded 296 place-names. These "large numbers alone do not account for the high frequency with which place-names typically appear in Western Apache discourse," he wrote. "In part, this pattern of recurrent use results from the fact that Apaches, who travel a great deal to and from their homes, regularly call on each other to describe their trips in detail. Almost invariably, and in marked contrast to comparable reports delivered by Anglos living at Cibecue, these descriptions focus as much on *where* events occurred as on the nature and consequences of the events themselves." This binding of specific experiences to specific locations, which is an effective way to remember both pieces of information reciprocally, also occurs in Western Apache storytelling. Place-names are "situating devices" in stories. "Thus, instead of describing these settings discursively, an Apache storyteller can simply employ their names, and Apache listeners, whether they have visited the sites or not, are able to imagine in some detail how they might appear."

Basso describes Western Apache culture as one in which storytelling, places, traveling, memory, and future imagining—all the elements involved in wayfinding—are the ingredients required for wisdom itself. One day Basso talked to Dudley Patterson, a horseman who lived in Cibecue, who tried to answer Basso's question: What is wisdom? "Your life is like a trail. You must be watchful as you go," Patterson told him.

> Wherever you go there is some kind of danger waiting to happen. You must be able to see it before it happens. . . . If your mind is not smooth you will fail to see danger. . . . If you make your mind smooth, you will have a long life. Your trail will extend a long way. You will be prepared for danger wherever you go. You will see it in your mind before it happens. How will you walk along this trail of wisdom? Well, you will go to many places. You must look at them closely. You must remember all of them. Your relatives will talk to you about them. You must remember everything they tell you. You must think about it. You must do this because no one can help you but yourself. If you do this your mind will become smooth. It will become steady and resilient. You will stay away from trouble. You will walk a long way and live a long time. Wisdom sits in places. It's like water that never dries up. You need to drink water to stay alive, don't you? Well, you also need to drink from places. You must remember everything about them.

When I read Patterson's words I felt how apt they were to all I had learned about the origins of the human mind in the practice of navigating across prehistoric landscapes, of our unique capacity to store memories and think in terms of stories. Life itself, I started to think, is a movement through time, the creation of a story of how we came to be here now and where we are going. It was much later that I came across the work of Patrick Henry Winston, a pioneer in artificial intelligence research who thinks that storytelling is *so* central to human intelligence, it is also the key to creating sentient machines in the future.

———

I met Winston at his second-floor office in the Stata Center on the MIT campus, a surreal 720,000-square-foot Alice in Wonderland building designed by Frank Gehry with walls and corners crashing into one another at sharp angles. Winston, white-haired and dressed casually, sat at his desk. Behind him was an assortment of books on the Civil War, of which he is an amateur scholar, but it was the painting above his head that caught my attention. It was a framed reproduction of Michelangelo's fresco *The Creation of Adam*, depicting the moment just before genesis, the fingers of God and Adam hanging in the air, about to touch and set into motion the biblical story of man on earth.

Winston has been at MIT for most of his life; he was an undergraduate there, receiving a degree in 1965 in electrical engineering before doing a doctoral dissertation under the famous artificial intelligence researcher and philosopher Marvin Minsky. It was Winston who took over the Artificial Intelligence Laboratory when Minsky eventually moved on to create the influential Media Lab. "I do screwball AI," Winston, now in his sixties, told me of his career. "My work is about developing a computational understanding of human intelligence rather than creating applied systems. Applied AI is about 95 percent of the field." Within this niche of screwball AI, Winston has created a new computational theory for human intelligence. He believes that in order to evolve AI beyond systems that can simply win at chess or *Jeopardy*, to build systems that might actually begin to approach the intelligence of a human child, scientists must tackle one of the age-old questions of philosophy: identifying what it is *exactly* that makes humans so smart. Logos? Imagination? Reason? It was Alan Turing, Winston explained to me, who published the seminal paper *Computer Machinery and Intelligence* in 1950 that argued human intelligence was the result of complex symbolic reasoning. Minsky also believed that reasoning—the ability to think in a multiplicity of ways that are hierarchical—was what made humans human. Subsequent AI researchers argued that human intelligence is a matter of genetic algorithms, statistical methods, or replicating the neural net of the human brain. "I think Turing and Minsky were wrong," said Winston, pausing. "We forgive them because they were smart and mathematicians, but like most mathematicians, they thought reasoning is the key, not the byproduct."

"My belief is that the distinguishing characteristic of humanity is this keystone ability to have descriptions with which we construct stories," he told me. "I think stories are what make us different from chimpanzees and Neanderthals. And if story understanding is really where it's at, we can't understand our intelligence until we understand *that* aspect of it." Winston draws on linguistics, in particular a hypothesis developed by fellow MIT professors Robert Berwick and Noam Chomsky, to explain how language evolved in humans. Their idea is that humans were the only species who evolved the cognitive ability to do something called "Merge." This linguistic "operation" is when a person takes two elements from a conceptual system—say "ate" and "apples"—and merges them into a single new object, which can then be merged with another object—say "Patrick," to form "Patrick ate apples"—and so on in an almost endlessly complex nesting of hierarchical concepts. This, they believe, is the central and universal characteristic of human language, present in almost everything we do. "We can construct these elaborate castles and stories in our head. No other animals do that," said Berwick. The theory flips the common explanation of why language developed: not as a tool for interpersonal communication but as an instrument of internal thought. Language, they argue, is not sound with meaning but meaning with sound.

In their book on the subject, *Why Only Us: Language and Evolution,* Berwick and Chomsky draw on brain imaging studies that have shown how the prefrontal cortex—a region important for language processing—has evolved. They propose that through encephalization, the evolutionary expansion and reorganization of our brains, a novel anatomical loop between the posterior superior temporal cortex (STC) to Broca's area (responsible for speech processing and coherent speech, respectively) was created. The maturation of these dorsal and ventral pathways in the brain's language and premotor areas during childhood enables each of us to do the merge operation and have symbolic language. Indeed when researchers have looked at the circuits of our brains that activate when the merge function is occurring, the process takes place in four different connecting tracks. Newborn babies, interestingly, aren't born with some of these connections in place, and studies show that, if this fiber tract in children is not fully matured between the posterior STC and Broca's area, the interpretation of syntactically complex sentences is poor. "They don't have the fat insulation, they're not wired up," suggested Berwick. "Over the course of a couple years [most children] are going to start talking, and that could be the result of a small evolutionary change. The brain gets bigger and the extra growth wired up these systems. The rest, they say, is history."

For Winston, the merge hypothesis represents the best explanation so far for how humans developed story understanding. But Winston also believes that the ability to create narrative stems from spatial navigation. "I do think much of our understanding comes from the physical world, and that involves things moving through it," he said. "The ability to put things in order, I think that comes out of spatial navigation. We benefit from many things that were already there, and sequencing was one of the things that was already there. . . . From an AI perspective, what merge gives you is the ability to build a symbolic description. We already had the ability to arrange things in sequences, and this new symbolic capability gave us the ability to have stories, listen to stories, tell stories, to combine two stories together to get a new story, to be creative." Winston calls it the Strong Story Hypothesis.

Winston decided to see if he could create a program that could understand a story. Not just read or process a story but glean lessons from it, even communicate its own insights about the motivations of its protagonists. What were the most basic functions that would be needed to give a machine this ability, and what would they reveal about human computation? Winston and his team decided to call their machine Genesis. They started to think about commonsense rules it would need to function. The first rule they created was deduction—the ability to derive a conclusion by reasoning. "We knew about deduction but didn't have anything else until we tried to create Genesis," Winston told me. "So far we have learned we need seven kinds of rules to handle the stories." For example, Genesis needs something they call the "censor rule" that means: if something is true, then something else can't be true. For instance, if a character is dead, the person cannot become happy.

When given a story, Genesis creates what is called a representational foundation: a graph that breaks the story down and connects its pieces through classification threads and case frames and expresses properties like relations, actions, and sequences. Then Genesis uses a simple search function to identify concept patterns that emerge from causal connections, in a sense reflecting on its first reading. Based on this process and the seven rule types, the program starts identifying themes and concepts that aren't explicitly stated in the text of the story. What fascinated Winston initially was that Genesis required a relatively small set of rule types in order to successfully engage in story understanding at a level that appears to approach human understanding. "We once thought that we would need a whole lot of representations," said Winston. "We now know that we can get away with just a few."

"Would you like a demonstration?" he asked me. I rolled my chair around to the other side of his desk and watched as Winston opened the Genesis program. "Everything in Genesis is in

English, the stories, the knowledge," he said. He typed a sentence into a text window in the program: "A bird flew to a tree." Below the text window I saw case frames listed. Genesis had identified the actor of the story as the bird, the action as fly, and the destination as tree. There was even a "trajectory" frame illustrating the sequence of action pictorially by showing an arrow hitting a vertical line. Then Winston changed the description to "A bird flew toward a tree." Now the arrow stopped short of the line.

"Now let's try *Macbeth*," Winston said. He opened up a written version of *Macbeth*, translated from Shakespearean language to simple English. Gone were the quotations and metaphors; the summarized storyline had been shrunk to about one hundred sentences and included only the character types and the sequence of events. In just a few seconds Genesis read the summary and then presented us with a visualization of the story. Winston calls such visualizations "elaboration graphs." At the top were some twenty boxes containing information such as "Lady Macbeth is Macbeth's wife" and "Macbeth murders Duncan." Below that were lines connecting to other boxes, connecting explicit and inferred elements of the story. What did Genesis think *Macbeth* was about? "Pyrrhic victory and revenge," it told us. None of these words appeared in the text of the story. Winston went back to a main navigation page and clicked on a box called "self-story." Now we saw, in a window called "Introspection," the process of Genesis's own understanding of the story, the sequence of its reasoning and inference. "I think that's cool because Genesis is a program that is in some respects self-aware," he said.

Building complex story understanding in machines could help us create better models for education, political systems, medicine, and urban planning. Imagine, for example, a machine that possessed not just a few dozen rules by which to understand a text but thousands of rules that it could apply to a text hundreds of pages long. Imagine such a machine employed by the FBI, given an intractable murder case with puzzling evidence and a multitude of potential perpetrators. Or inside the Situation Room, offering American diplomats and military intelligence its own analysis of the motives of Russian hackers or Chinese belligerence in the South China Sea, calculating predictions for future behavior based on an analysis of a hundred years of history.

Winston and his students have used Genesis to analyze a 2007 cyberwar between Estonia and Russia. They have also found creative ways to test its intelligence, prompting it to tell stories itself, or tweaking its perspective to read a story from different psychological profiles—Asian vs. European, for example. One of Winston's graduate students gave Genesis the ability to teach and persuade readers. For example, the student requested that Genesis make the woodcutter look good in the story of Hansel and Gretel. Genesis responded by adding sentences that emphasized the character's virtuousness.

Recently, Winston's students found a way to give Genesis schizophrenia. "We think some aspects of schizophrenia are the consequence of fundamentally broken story systems," explained Winston. He showed me a cartoon illustration. It depicted a little girl trying to open a door handle that is too high to reach, and then fetching an umbrella. A healthy person would infer the girl is getting the umbrella to extend her reach and open the door; a person with schizophrenia does what is called hyperpresumption—inferring that the girl is getting an umbrella to go outside in the rain. To get Genesis to think like a schizophrenic in this way, Winston and his students simply switched two lines of code in the program. They put Genesis's search for explanations that tie story elements together after the search for the default answer, the one that the girl will be going into the rain. Hyperpresumption, according to Genesis, is a dysfunction of sequencing in the brain. They called it the Faulty Story Mechanism Corollary.

One of Winston's students at MIT was Wolfgang Victor Hayden Yarlott, an engineering and computer science major who graduated in 2014 and is now doing a PhD at Florida International University. Yarlott is a Crow Indian and had an idea: if Winston was correct about the Strong Story Hypothesis, that stories are a key part of human intelligence, Genesis needed to demonstrate that it understood stories from all cultures, including indigenous cultures such as the Crow. "Stories are how intelligence and knowledge is represented and passed down from generation to generation— an inability to understand any culture's stories means that either the hypotheses are wrong or the Genesis system needs more work," Yarlott wrote in his thesis.

Yarlott chose a set of five Crow stories for Genesis to read, including creation myths that he'd heard during his childhood in southern Montana. His challenge was to create recognition in Genesis of chains of events that seemed unrelated, of supernatural concepts like medicine (which has a magiclike quality in Crow folklore), and of the "trickster" personality traits. Those are all elements in the Crow stories that, as Yarlott determined, distinguish them from the canon of English-European storytelling. The creation myth "Old Man Coyote Makes the World" features animals that communicate with Old Man Coyote just as people do. As Yarlott points out, there is an amazing display of power or medicine that enables Old Man Coyote to create—but there are explicitly unknowable events that take place as part of the story, like, "How he did this, no-one can imagine." To solve these issues Yarlott had to give Genesis new concept patterns to recognize. For example,

> Start description of "Creation".
>   XX and YY are entities.
>   YY's not existing leads to XX's creating YY.
> The end.
> Start description of "Successful trickster".
>   XX is a person.

```
    YY is an entity.
    XX's wanting to fool YY leads to XX's fooling YY.
The end.
Start description of "Vision Quest".
    XX is a person.
    YY is a place.
    XX's traveling to YY leads to XX's having a vision.
The end.
```

The stories that Yarlott told to Genesis read like this:

```
Start experiment.
Note that "Old Man Coyote" is a name.
Note that "Little_Duck" is a name.
Note that "Big_Duck" is a name.
Note that "Cirape" is a name.
Insert file Crow commonsense knowledge.
Insert file Crow reflective knowledge.
"Trickster" is a kind of personality trait.
Start story titled "Old Man Coyote Makes the World".
Old Man Coyote is a person.
Little_Duck is a duck.
Big_Duck is a duck.
Cirape is a coyote.
Mud is an object.
"the tradition of wife stealing" is a thing.
Old Man Coyote saw emptiness because the world didn't
    exist.
Old Man Coyote doesn't want emptiness.
Old Man Coyote tries to get rid of emptiness.
```

Yarlott found that Genesis was capable of making dozens of inferences about the story and several discoveries too. It triggered concept patterns for ideas that weren't explicitly stated in the story, recognizing the themes of violated belief, origin story, medicine man, and creation. It seemed to comprehend the elements of Crow literature, from unknowable events to the concept of medicine to the uniform treatment of all beings and the idea of differences as a source of strength. "I believe this is a solid step towards showing both that Genesis is capable of handling stories from Crow literature," he surmised, "and that Genesis is a global system for story understanding, regardless of the culture the stories come from."

Genesis has obvious shortcomings: so far it only understands elementary language stripped of metaphor, dialogue, complex expression, and quotation. To grow its capacity for understanding, Genesis needs more concept patterns—in other words, more teaching. How many thousands of stories does a child hear, create, and read as she grows into adulthood? Perhaps hundreds of thousands.

But there are probably fundamental limitations to the machine's potential. One rainy day in November I went to see Winston teach his extremely popular undergraduate course at MIT, "Intro to Artificial Intelligence." I listened to him explain the Strong Story Hypothesis to hundreds of students and demonstrate its abilities. "Can Watson do this?" he quipped. But then he posed a series of questions to his young students casting doubt on his own invention.

"Can we think without language?"

The room was quiet.

"Well, we know from those whose language cortex is gone that they can't read or speak or understand spoken language," he explained. "Are they stupid? They can still play chess, do arithmetic, find hiding places, process music. Even if the external language apparatus is blown away, I think they still have inner language."

He paused. "Can we think without a body?"

Quiet.

"What does Genesis know about love if it doesn't have a hormonal system?" he said. "What does it know about dying if it doesn't have a body that rots? Can it still be intelligent?"

He paused again. "We'll leave that part of the story for another time."

Love and death, I marveled, as students filed out of the lecture hall. The stuff of the most epic stories on earth. What could Genesis understand of these universal human conditions without embodiment in time and space?

PART TWO

—

AUSTRALIA

SUPERNOMADS

Wherever I went in Australia, I wondered if the road I traveled on followed a Dreaming track—the network of Aboriginal trade routes and cultural thruways that crisscrosses the whole of the continent like a noospheric highway system. In Aboriginal cosmology, the Dreaming is a period of history during which their ancestors took animal forms and created the topography of the land by traveling and leaving behind tracks in the earth. The features of the land are the evidence of their journeys and are also called *story-strings* or *songlines*. Dreaming tracks aren't etched on the land. They live in the memories of individuals who inherited the routes from the previous generations, who inherited them as well, creating one of the oldest chains of human memory in history. Driving at seventy miles per hour along the concrete roads of Australia's modern landscape, I wondered whether I was treading on the routes of individuals who had walked this way to get to families, ceremonies, trade fairs, harvests, or sacred sites. What stories and songs had helped them remember the way to go?

Before contact with European colonizers, Dreaming tracks were conduits for informational exchange between different Aboriginal nations and communities. One community might have part of the story that runs to the border of their territory, and that story continues on into another territory. In other words, stories are instruments by which the history, geography, and laws of a group and other groups can be learned and shared orally. Anthropologists recorded Aboriginal people who knew about the intricate tribal relationships of those living over a thousand miles away. A single epic story could cross languages: the Fire Story connects the people of Wangkangurru and Wangkamadla with the Diamantina and Georgina Rivers and then the Arrernte people. In 2012, Putuparri Tom Lawford, a lawman of the Wangkajunga people, described as part of an oral history project how storylines link people in the desert. "Well some other tribes, some storyline or songline they cut through that tribe and through other tribes too," he said. "You know this songline comes from that area, through this area, cuts through and finishes in this mob area here. That song itself will tell you. When they are singing a song, it's a story, it will tell you how far it comes from this tribe to another tribe. And that is the good thing about all Western Desert people, is that we got the one songline that follows on. Even though we come from different parts of the Great Sandy Desert."

When British ships arrived in 1788, the cartography of colonization and Dreaming tracks began to overlap. Western explorers and stockmen used Aboriginal scouts and their knowledge of the landscape and watering sites to lead the "charge into the spatial geography of the new lands." As a result, new roads often followed the Aboriginal trading pathways, and even some railways were created alongside Dreaming tracks. Today, parts of highways in the Bunya Mountains in Queensland overlap with the Euahlayi people's star maps, sequences of stars in the night sky whose appearance represents waypoints for the traveler to follow along a route. A stretch of the Victoria Highway in the Northern Territory follows along a Dreaming track of the Wardaman people. Most famously, the Canning Stock Route, a thousand-mile passage through the Western Desert that crosses some fifteen different Aboriginal language groups, overlays and intersects with myriad Dreaming routes.

The history of the Canning Stock Route also encapsulates the horrors of European brutality toward Aboriginal people and the galactic differences in how the two cultures approached geography. For Aboriginal people, "the whole Western desert is criss-crossed with the meandering tracks of ancestral beings," wrote social anthropologist Ronald Berndt, "mostly though not invariably following the known permanent and impermanent waterhole routes." In the early 1900s Alfred Canning decided to create a route to bring beef cattle to market across the same desert, forging a lucrative economic artery through the country. He knew that the endeavor was doomed without Aboriginal knowledge of water soaks and springs to sustain animals and their masters. So he forced the Aboriginal scouts, men by the names of Charlie, Gabbi, Bandicoot, Politician, Bungarra, Smiler, Sandow, and Tommy, to give up their knowledge by taking them hostage. Canning used neck chains and handcuffs so that the scouts couldn't escape, and in the day he fed them salted beef in order to increase their thirst for fresh water. Each time they entered new country and encountered a member of a different group, Canning would "try and get him before we let the other go. Then the native we had would speak to the new one. They will come in willingly with another native. We would then get him as soon as possible to draw a plan on the ground of the different waters."

At each of these sites, Canning later built a well, deepening the soaks and springs and building walls down into the earth. What he likely didn't know or care about was that the sites—and the entire landscape he traveled through—had been created by his scouts' Aboriginal ancestors back in the Dreamtime. There are few aspects of Australia that the ancestors are not responsible for; they created rivers, waterholes, rocks, valleys, and hills. Most features have a story associated with their creation, told in song during religious ceremonies and travel. As the anthropologist Deborah

Bird Rose has described, the Dreaming is forever and exists everywhere; it cannot be changed or washed away with time. "The earth is the repository of blood from Dreaming deaths and births, sexual excretions from Dreaming activities, charcoal and ashes from their fires," she writes in *Dingo Makes Us Human*. "Dreaming life has this quality which defies change: those things which come from Dreaming—country, boundaries, Law, relationships, the conditions of human life—endure."

One drawing by a Martu elder of the Canning Route shows songlines on an east-west axis intersecting the route at every stage. "There are wells on the Canning Stock Route but they are people's water," explains Lawford. "And in a way Alfred Canning, he trespassed onto people's land, Country. He took over their waters for animals, to feed cattle. . . . Well the Canning Stock Route, it broke the Country up."

The Dreaming is not easy for non-Aboriginal people to conceptually grasp. Explanations in English often sound oxymoronic (it is in the past but has no end, for instance), or like the romantic imaginings of New Age primitivists. The word "Dreaming" is merely adequate at communicating its substance; it was first used in the nineteenth century by a postmaster and ethnologist in Alice Springs, Francis Gillen, and was later popularized by the biologist and anthropologist Walter Baldwin Spencer. Gillen was trying to translate an Arrernte word pertaining to reality and religion. In 1956, anthropologist W. E. H. Stanner attempted to coin a different word, the "everywhen," but "Dreaming" is the word that stuck. Robyn Davidson, who famously traveled across the Australian desert by camel as a young woman, has called the Dreaming the original "theory of everything," but one that she never could fully comprehend. "No matter how much I read about the Dreaming, the confidence that I understand it never quite takes root in my mind," she wrote in an essay about twenty-first-century nomads. "To me it is on a par with, say, quantum mechanics, or string theory—ideas you think you grasp until you have to explain them." Despite this, Davidson sums it up as "a spiritual realm which saturates the visible world with meaning; that it is the matrix of being; that it was the time of creation; that it is a parallel universe which may be contacted via the ritual performance of song, dance and painting; that it is a network of stories of mythological heroes—the forerunners and creators of contemporary man."

The reason I went to Australia was to understand whether Dreaming tracks were a mnemonic device for wayfinding. To me it seemed that Dreaming tracks and their song-cycles are literal directions for a traveler to follow. Aboriginal people seemed to have created cultural traditions that took advantage of the human mind's narrative proclivities. By associating stories with specific places and encoding navigation information within the sequences of songs or stories, they made it easier to recall them through reciting these oral maps. This strategy is not unlike the Greek memory palaces, except Aboriginal people treated the *landscape* as a memory palace rather than inventing imaginary ones. As David Turnbull writes in his book *Maps Are Territories*, "Thus the landscape, knowledge, story, song, graphic representation and social relations all mutually interact, forming one cohesive knowledge network. In this sense, given that knowledge and landscape both structure and constitute each other, the map metaphor is entirely apposite. The landscape and knowledge are one as maps, all are constituted through spatial connectivity."

Until the 1960s, it was widely accepted in anthropology that the Australian continent had been inhabited for a mere ten thousand years, less than both North and South America. In the 1990s anthropologists began reexamining ancient skeletal remains, and with improved dating techniques they discovered that the first Australians arrived on the continent *at least* forty thousand and perhaps as long as seventy thousand years ago. The first Australians were likely a founding population of around a thousand that would have arrived in small groups by boats from the north, possibly Timor or New Guinea, which was at one point separated from Australia by as little as fifty-five miles of ocean. The archaeologist Scott Cane describes this early history of ancient Aboriginal people in his book *First Footprints*; how their arrival on the continent coincided with an explosion of human migration around the globe, and in just over three thousand years, humans occupied every continent from Africa to Australia. By around fifty thousand years ago, humans had reached central southeast Australia, and by forty-four thousand years ago had reached Tasmania (once connected to the mainland by an ice bridge). Cane refers to these early settlers as "super-nomads," a people equipped with endurance, physical skill, precise navigation abilities, and, he ventures, an encyclopedic knowledge of nature. By the time the first European colonizers "discovered" Australia in the seventeenth century, there were some 250 languages spoken there and a population of perhaps a million people. The origin stories told by many Aboriginal groups are similar to those that geneticists have reconstructed from DNA samples: the first landing occurred in the north and spread downward across the continent. In many stories, this is described as a mother emerging from the ocean and onto dry land. In some she carried a dilly bag full of babies and a walking stick, and as she traveled she planted babies in the land and made holes with her stick that filled with water.

The anthropologist Deborah Bird Rose spent two years among the Yarralin people of the Victoria River Valley and found that there is a temporal distinction between Dreaming and ordinary time. Most senior people, she wrote, trace their genealogies back three generations and talk about their grandparents coming from the Dreaming; ordinary time begins about a hundred years ago and is made up of the day-to-day passage of time, marked with things like aging and the changing of seasons. This means that Dreaming is forever, but also precedes us. Rose conceptualized this seeming contradiction through the image of a great wave that "follows behind us obliterating the debris of our existence and illuminating, as a synchronous set of events, those things which endure."

The Dreaming tracks enabled a thriving economic marketplace in Aboriginal society, something that European colonizers thought was impossible. They were convinced that Aboriginal people wandered the land in a struggle for survival and a constant search for food and water. That perception of an impoverished race on the brink of extinction lingered for many hundreds of years, but today it is irrefutable that Aboriginal trading routes could easily compete with civilization's greatest examples, such as the Inca and Incense roads. As Dale Kerwin, an Aboriginal historian and scholar at Griffith University, has extensively documented in his book *Aboriginal Dreaming Paths and Trading Routes*, Aboriginal people traded red ochre, pearls, spears, baskets, fishhooks, nuts, grinding stones, axes, boomerangs, resin, and intellectual property—songs containing information on how to find waterholes and food—over thousands of miles and for eons. One of the mines where red ochre was collected for trade was in continuous use for 20,000 years; the Silk Road only began some 2,200 years ago and lasted about 1,600 years. Some of the longest Aboriginal trading routes were created for pituri, a native nicotine plant. Cultivated and then baked in sand ovens, it could be chewed or smoked and acted as a stimulant; it was traded over an area of at least 340,000 square miles. At times as many as five hundred traders would gather to sell and buy it. Pituri was moved along a virtual "highway," a Dreaming track that stretches for some 2,300 miles and is made up of stories with regional variants. For example, from Port Augusta to Alice Springs the story is the Urumbula or Native Cat Dreaming, and across the Simpson Desert it is known as a Two Dogs' Dreaming story. Kerwin cites an Arrernte elder, Isabel Tarrago, from the Simpson Desert, whose mother cultivated pituri and was a "song woman" who knew the trade routes. "We are linked by song to people at Borroloola by the Dog Dreaming, the mob [group] there are connected to granny and mum through this song and so am I," said Tarrago. "We are related across huge distances by extension of this Dreaming and song. The country is the text to be read and song is the means to unravel the text."

Kerwin has written that Dreaming tracks were an aid for spatial orientation, and that a further aid "resides in mnemonics and rote learning associated with oral traditions and songlines. This orientation technique relies on the ability of the traveller to recall ideas and experiences that have been imprinted into social memory." The Aboriginal traveler could implicitly trust the logic and navigational skills of the ancestors who traveled before them. Kerwin explains how the ancestors employed tremendous wisdom when they created the Dreaming tracks:

> [The ancestors] know on which side an obstacle can be passed, where there is firm land which leads through a bog, and where the best going is: sand, rock, or dry soil. The spirits know the easiest approach. And through their work and through infinite time, the Dreamtime spirits sculptured the landscape and taught how the country should be read. Aboriginal people see the country in the landscape and the Dreamtime paths everywhere.

In some cases, the mistakes of ancestors are manifest in the landscape and linger as a lesson for the traveler. For the Arrernte people, gum trees around the Todd River are the Caterpillar people that were trying to reach Emily Gap from Mt. Zeil and became lost; they remain there as evidence of how *not* to get to Emily Gap. Often songlines tell a traveler where a waterhole is, or a claypan that might represent the former camp of an ancestor and indicate the presence of a certain animal or bush vegetable. In this way the songs are a kind of memory device, investing emotion and meaning into the earth by turning even the most innocuous-looking rock or hill into a story of how it came to be and why it looks the way it does. In other words, the stories create landmarks for the mind to hold onto. One term for these memory aids, which arguably increased survivability in the bush, is "totemic geography," which, as anthropologist Luise Hercus has said, gives "deeper significance to ordinary geography and makes it more memorable."

When was the first Dreaming story told? There is no way to know precisely. But even the most conservative estimate makes Aboriginal oral history the oldest in the world. Until very recently, there was a consensus that the longest time period that human memories can be transmitted between generations before their meaning has completely changed or become obscured from the original is five hundred to eight hundred years. But in 2016, two Australian researchers published a paper in the journal *Australian Geographer* that upended this idea. Patrick Nunn and Nicholas Reid recorded stories from twenty-one locations around coastal Australia, from the Gulf of Carpentaria in the north to Kangaroo Island in the south. In each place they found stories about a time when parts of the coastline now under the ocean were actually dry land. The researchers matched the stories to geological evidence of post-glacial sea-level rise. It seems that these stories have been repeated from one generation to the next for a minimum of seven thousand years but possibly for as long as thirteen thousand years and represent "some of the world's earliest extant human memories."

Nunn and Reid explain these nearly unbelievable figures by pointing to several characteristics of Aboriginal culture that elucidate how such a faithful oral transmission was made possible. They point to the great value Aboriginal people place on precision in telling stories "right." Additionally, not everyone has the authority to tell a story; only some people have ownership over them and are therefore held accountable to learn them faithfully. "For example, a man teaches the stories of his country to his children," write Nunn and Reid. "His son has his knowledge of those stories judged by his sister's children—for certain kin are explicitly tasked with ensuring that those stories are learned and recounted properly—and people take those responsibilities seriously."

By embedding story in landscapes and the actions of ancestor beings, the Aboriginal oral tradition of songlines is strikingly similar to other oral traditions—the ballads, epics, and children's rhymes found in places like Ireland, Yugoslavia, ancient Greece, and numerous folk traditions. In his book *Memory in Oral Traditions*, the cognitive neuroscientist David Rubin writes that there are certain properties that all these oral traditions share. Orally transmitted poems and epics are concrete, that is, they use actions by agents, like heroes or gods, who perform specific acts that are easy to visualize. The subjects of epic poems and songs are rarely abstract concepts like justice or heroism. Instead, these qualities are illustrated through the actions of protagonists. Second, Rubin argues, oral epics are almost always spatial because this makes it easier for the human mind to remember them. "There are no one-scene epics; travel is the rule," writes Rubin. "Homeric epic refers to itself as *oime* (path). The *Odyssey* is an odyssey, with hypothetical maps of travel shown in some editions. Even in the *Iliad*, where large segments occur between Troy and the sea, the location of battles and other events is constantly changing. Because such spatial layout typically follows a known path, order information can be preserved. In contrast, it might be confused if there were one large, simultaneous image."

Rubin thinks that oral traditions developed to avoid weaknesses of human memory, which more easily records scenes rather than abstract knowledge. And these traditions utilize another strength of our brains: using rhythm and music to cue memory. Consider how many of us learn the alphabet as a child by singing it. With some practice, the notes become bound to the letters, allowing the mind to recall them with ease. Same with other popular children's verses or songs, like "Row, Row, Row Your Boat" or "Mary Had a Little Lamb." In some cases, it's difficult to extract the words from the melody. Because in our memory, the music and the words are inseparable. Gregorian chanting used music as a mnemonic device by pairing psalms and mass chants to melodies; one academic estimates that the Gregorian chant repertoire included nearly four thousand texts by the Middle Ages. Rubin has argued that we shouldn't think of memory in its most common sense, as abstract traces in the mind, but as a socially guided rhythm of body movement and gesture that is an integral part of transmission. In oral cultures, the kind of rhythm used to memorize songs and the information contained within them exists "only in its performance, and that performance is as much motor as it is verbal."

Rubin's hero was a Harvard literature professor by the name of Albert Lord, who in 1960 published *The Singer of Tales*, a canonical text on oral traditions. For years, Rubin made summer pilgrimages to visit Lord, and he told me, "Lord did not even like the word *memory*. No one memorizes an epic poem. They *sing* it."

DREAMTIME CARTOGRAPHY

One morning I took a taxi from a sleepy suburb south of the city of Perth to the airport and boarded a four-hour plane ride to Darwin, Australia's tropical northernmost settlement, which sits on a nub jutting into the Timor Sea. My window seat faced west, and for most of the time we flew over the Western Desert, a region of some five hundred thousand square miles that includes the Gibson, Great Sandy, and Great Victoria Deserts. One European explorer referred to this land as a "vast howling wilderness." To me, the landscape looked like watercolor paints spilled on parchment and then torched by heat. Pink, red, and ochre earth swirled together; venae cavae of ancient riverbeds and salt lakes that appeared shrunken and concentrated in dark purples and scalding whites. Its size scrambled my mind. Wherever they went in the "New World," European colonists were convinced they had discovered *terra nullius*, "land belonging to no one," which they eagerly laid claim to. Australia was the same. "Where, we ask, is the man endowed with even a modicum of reasoning powers, who will assert that this great continent was ever intended by the Creator to remain an unproductive wilderness?" asked the *Sydney Herald* in 1838. From thirty thousand feet, central Australia still looks empty; there are no cities or agriculture projects evincing human dominion of the desert. But *terra nullius* was, and is, a lie. The eye of a foreigner can still see this desert as a vast howling wilderness, when in fact the Dreaming—called "Tjukurrpa" by the Pitjantjatjara people of central Australia—and the tracks left by ancestors in the desert represent tens of thousands of years of uninterrupted human habitation.

Outside Perth, our flight path took us northeast toward the town of Meekatharra, not far from the start of the Canning Stock Route. I thought about how the Aboriginal scouts in Canning's party seemed to have taken enormous pains to not give up the location of the most secret Dreaming places: the stock route twists and turns even as it slices through the land on a north-south axis. "Whitefellas just reckon go where the straight line is," said Jawurji Mervyn Street, a cattleman and artist who lives in the Yiyili community in the Western Desert, about the route. "In the Martu side there's no straight line. You can't go straight when you got some special thing in the road. You're gonna have to dodge around. The stock road circles round and round, and I been thinking straight away: might be some special place there, and the guides made it clear all the way by going around it." Despite the constant threat of violence, the scouts tried to protect the sanctity of the country.

About midway through the trip, the plane passed over the Simpson Desert, the place where the doctor and explorer David Lewis traveled with two Antikarinya men, Wintinna Mick and Mick Stewart, back in 1972. Their journey together represents one of the first cases in which someone attempted to record Aboriginal navigation practices, and as a result discovered that the people wayfinding in Australia's deserts possessed some of the most precise orienting abilities known to man.

Twelve years before he went to Australia, Lewis had participated in the world's first single-handed transatlantic yacht race, and he later circumnavigated the world with his wife and two young daughters. Born in Britain but raised in New Zealand, he was fascinated by traditional South Pacific navigation and undertook a sailing trip from Tahiti to New Zealand without using a compass or sextant. His book, *We, the Navigators* (1972), was the summation of eight years of research into Polynesian and Micronesian navigation traditions, including sailing thirteen thousand miles in the western Pacific and the first circumnavigation of the world in a catamaran. Always eager for the next adventure, Lewis went to Australia because he had become fascinated by Aboriginal route-finding and launched a three-year research project to study spatial orientation practices.

Once there, however, Lewis discovered he was unprepared for what he encountered. While he assumed he would find individuals who used environmental guides like the sun and stars for orientation, much like the South Pacific Islanders he had previously written about, he instead found something entirely different. "While there are certain elements very roughly analogous to the mould I had in mind of non-instrumental maritime usage," described Lewis, "this was a misleading formulation on two counts, since it tended to convey connotations of featureless landscape and of a celestially based system of reference." Lewis realized that in Australia, no such thing as a "featureless landscape" existed. Furthermore, Aboriginal spatial orientation didn't seem to fit into any known theory of navigation that Lewis was aware of. Mick and Stewart, his companions on his first foray into the desert, never seemed to use external references of any kind save for a few intermittent and, what seemed to Lewis, unremarkable landmarks. Instead, what they did was take a mental note of the starting point of a trip and then travel blindly through the landscape until the hill, waterhole, tree, or rock they were aiming for appeared. They rarely took stock of their route or reevaluated their direction, yet they could travel great distances with unerring accuracy; it was almost as though Mick and Stewart had simply managed to memorize hundreds of square miles of desert landscape. Lewis began to wonder whether Mick and Stewart might possess a total recall of every topographical feature of any country they had ever traveled in.

"A single visit 40 years ago would be sufficient to make an indelible imprint," he wrote. When Lewis asked Mick to explain how he knew which direction to travel in, Mick just told him, "I have a feeling." Then he drew Lewis a map in the sand to demonstrate. "If I go south 10 miles, then a little east, to get back home I must go north 10 miles and a little west," he said. "If there are no landmarks I still know the directions. Aborigines knew north, south, east and west before the white man's compass." Okay, persisted Lewis, but how had Mick found his way over *twenty-three miles* of identical sand hills that day? "I know this north-west direction," said Mick, "not by the sun but by the map inside my head."

Over the next three years Lewis worked as a visiting fellow at the Australian National University in Canberra and traveled nearly five thousand miles by Land Rover and on foot all over the Simpson Desert and along the Canning Stock Route. He worked with Antikarinyha, Pintupi, and Luritja language speakers, asking them to point to unseen places after traveling and then comparing the results to his compass readings. Again and again, their accuracy was unerring.

One day in Western Luritja-Aranda country, two Pintupi men by the name of Jeffrey Tjangala and Yapa Yapa Tjangala brought Lewis to what seemed like a featureless landscape of squat mulga trees and spear grass. As he later described in the journal *Oceania*, the earth was flat and devoid of big trees, creeks, and sand hills, and visibility was limited to about three hundred feet. The men saw a *malu* (kangaroo) and stopped the Land Rover to shoot it, but the .22 bullet only wounded the animal. In order to track it through the bush, they left the vehicle behind and walked into the mulga. After half an hour on foot, they killed the kangaroo, and then Jeffrey and Yapa Yapa started to make their way back. "How do you know we are heading straight towards the Land Rover?" Lewis asked. Jeffrey touched his forehead and then swung his arm around to show how they had followed the *malu* round one way and then another. "We take a short cut," he said. "Are you using the sun?" asked Lewis. "No," said Jeffrey, and then walked for fifteen minutes directly to the car.

Lewis wrote that the "Pintupi's route-finding by these unremarkable landmarks was uncannily accurate. They alway knew just where they were, they knew the direction of spiritually important places for hundreds of kilometres around, and were oriented in compass terms." One night Jeffrey drew the cardinal directions in the sand. "North, south, east and west are like this in my head," he told Lewis. Then he pointed out the directions of all the significant Dreaming sites between their camp and his home near Lake Disappointment 250 miles away. It was then that Lewis realized that the spiritual world, the sacred sites and Dreaming tracks, were Jeffrey's primary references for orienting, and he experienced awe, fear, love, and extreme attachment to this spiritual geography. Lewis wondered whether he would ever be able to look at a hill or a rock hole with the same eyes as before. "All my preconceived ideas about 'land navigation' turned out to be wrong," he wrote. "In place of the stars, sun, winds and waves that guide Pacific Island canoemen, the main references of the Aborigines proved to be the meandering tracks of the ancestral Dreamtime beings that form a network over the whole Western Desert."

In the desert, Aboriginal people needing to communicate directions might make a drawing in the mud or sand. It begins with a circle that could represent water, a rock hole, a sacred site, a fire, or where the person was conceived, born, or initiated to his Dreaming. From this starting point, the cartographer draws a line representing a day's journey on foot, somewhere from three to ten miles, perhaps more. Then another circle—water, a landmark, another place of Dreaming, the next event in a creator-hero's story. Thus the map would appear in the very earth it represented, a network of circles and lines depicting an abstracted topography of history, cosmology, lived experience, and geography. Confronted with this symbolic representation that seems to collapse cartography, myth, and art, the outsider might be forgiven for disputing its wayfinding utility. "Often it is difficult for a Western European to enter their world," wrote the anthropologist Norman Tindale. "We are trained on cartographic plans with a compass as aid and relatively accurate determinations as to angle and distance, and by writing in place names and using symbols we share and use a relatively uniform series of conventions."

Tindale is credited as one of the first white Australians to understand that Aboriginal people were not starving wanderers in an endless search for food and water, when both the academic establishment and the Australian government were invested in the notion that indigenous people had no established territories and therefore no ownership of the land. In 1921, Tindale, a shaggy-haired young man barely into his twenties, spent a year on the island of Groote Eylandt in northeastern Australia. At the time, it was the longest period a scientist had ever spent with an Aboriginal community. One of the people he often worked with was Maroadunei, a Ngandi man who described to Tindale the features of his land and the borders between different language groups, how when one community's Dreaming stories ended, another's picked up. Tindale realized that Aboriginal people inhabited distinct territories from one another. He went home and drew a map of the boundaries between different groups. But when he went to publish the map and his research, his boss at the Australian National University told him to remove all the borders. It was almost two decades before Tindale's map, *Aboriginal Tribes of Australia,* was published in 1940. It was, in the words of one colleague, "radical in its fundamental implication that Australia was not *terra nullius.*"

Throughout the nineteenth and twentieth centuries, the Australian government moved Aboriginal communities onto reserves, cattle stations, missions, or into towns and cities in order to convert land for ranching and mining. The process of moving from what Aboriginal people knew as "Country" and the whites deemed to be wilderness was known as "coming in." Some of David Lewis's traveling companions—mainly those from the Pintupi language group—had been born and initiated in the desert and were considered among the last Aboriginal people to come in from the bush. Their departure from the desert began in the 1950s when the British started testing rockets and cleared the Pintupi from the flight path.

In 1962 Freddy West Tjakamarra, one of Lewis's traveling companions, left the desert for the town of Papunya, 155 miles west of Alice Springs, bringing his family and a group of trackers, including Nosepeg Tjupurrula. To get to Papunya, they walked several hundred miles from a rock hole site known as Mantati. "A lot of kids run up to meet us," remembered Bobby West Tjupurrula, Freddy's young son at the time, about their arrival in the settlement. "I was a bit excited and I felt ashamed coming to the big community, so I got my little spear and my little woomera [spear thrower]! . . . I wanted to go to school and learn. It is good, so I keep coming to school every day. We used to go hunting— walkabout—look for kangaroos, goanna, everything, every Friday afternoon, Saturday, Sunday and come back to school. We loved going hunting, camping."

It was at school that Bobby West witnessed a revolutionary movement in Aboriginal history. In June 1971, a group of Pintupi men painted a Dreaming story on the wall of the school. "I was a boy, teenage," described West. "Everyone come together and start telling a story, and what we're going to paint. One old man got up, said, 'We're going to draw this one—Honey Ants.' They come together and start helping him paint. It was very important, the mural, because they're proud to tell the story."

The honey ant mural depicted the Dreaming story of ants who later became men and the convergence of songlines at Papunya. Though painted over a few years later, the act of creating the mural is now considered a spark that led to an explosion of creativity known as the Western Desert Art Movement, in which Aborigi-

nal men—and soon after, women—formed cooperatives and began painting contemporary art depicting the Dreaming. Many of
the paintings were modified to protect sacred secrets, but the
act of painting itself was an expression of profound resiliency
and Aboriginal defiance after decades of colonial subjugation
that severed their connection to Country. The paintings were
proof of their relationship to the land, their intimate knowledge of its creation; in some later land claim cases, the paintings
were used as legal documents to prove generational use and
ownership in the courts. The act of creating the paintings struck
me as similar to the act of singing or traveling along the Dreaming tracks, an act of devotion that kept their relationship to the
land alive.

David Lewis unknowingly walked into the epicenter of this nascent movement when he arrived in Papunya in 1972. And many
of the men who traveled with him into the desert to demonstrate
their wayfinding practices—Long Jack Phillipus Tjakamarra,
Yapa Yapa Tjangala, Freddy West Tjakamarra, and Anatjari
Tjampitjinpa—were part of the now-legendary group that gave
birth to the Western Desert Art Movement. Of the eleven original shareholders of the Papunya Tula Artists Pty. Ltd., the first
artist collective, seven traveled with Lewis over three years so he
could understand how they navigated. One of them, Billy Stockman Tjapaltjarri, a petite man with a broad brow and deep-set
eyes who painted exuberant works that would be compared to
Jasper Johns and Joan Miró, accompanied Lewis in the Western
Desert in 1973. Uta Uta Tjangala and Nosepeg Tjupurrula, considered masters of Aboriginal contemporary art, took Lewis to the
Canning Stock Route in 1974. All had participated in the honey
ant mural, and in the early 1970s they were teaching children at
the Papunya school about painting and stories while depicting
and interpreting their own Tjukurrpa stories with paint on flat,
two-dimensional canvases for the first time.

For Westerners, these paintings are reminiscent of Wassily Kandinsky, Salvador Dalí, or Pablo Picasso. They have a mesmerizing, surreal geometry that the viewer is compelled to try and decipher. In one of Uta Uta Tjangala's earliest works, *Medicine Story 1971,* he depicts the story of a sorcerer in rich plum and mustard acrylics. Two phallic ovals are surrounded by circles and lines showing the journey from Ngurrapalangu, the same place in the Gibson Desert where Uta Uta was conceived around 1926, to Yumari, where the sorcerer has illicit sex with his mother-in-law. The painting shows the Old Man's testicles connecting to the ovals with mustard lines. These are life-giving waterholes and the paths between them. In 1974, Uta Uta revisited the Dreaming stories of Ngurrapalangu in a painting of the same name, depicting the story of two women and Short Legs, who fled from Old Man toward Wilkinkarra. The women created a claypan, where food grows after the rainfall, from their dancing, while Short Legs crawled into a cave and displaced the sacred objects that then became hills.

When I began to read about David Lewis's journeys into the Simpson Desert in the 1970s, I recognized several of the names of his companions as the same artists whose works I'd seen on the walls of museums—which sometimes sell for hundreds of thousands of dollars at auction. I didn't think that this connection between Lewis's navigation research and the art movement was coincidence. Lewis sought out men who were expert trackers and hunters, who had been born and initiated in the desert and knew its topography by heart. The same reverence and insatiable interest that made them master navigators—their intimate relationship to the Dreaming and their encyclopedic knowledge of bushcraft—also seemed to make them creative sages.

Whenever Lewis went into the desert with the men from Papunya, he was surprised by their passion for traveling. "I failed fully to understand the deep satisfaction elicited in my Aboriginal friends by monotonous driving from dawn to dusk day after day across a landscape that was vivified in sacred myth," he wrote.

"Every terrestrial feature, plant or track of an animal was meticulously noted and aroused very lively discussion. Highly coloured subsequent accounts of the features of the country traversed, such as the height of the sandhills, the colour of the rocks, the profusion of honey flowers, were given to envious friends back at the settlement."

"Maps" is arguably too limited a term for the complex layering of metaphor and history in modern Aboriginal art. But no one can deny the paintings' direct connection to the terrestrial geography of place. The paintings are of the topography of the Dreaming, and the Dreaming is the sacred geography of the land. "Aboriginal roads and tracks are maps; they are connections to country, they are about human movement, metaphor journeys, and the link between the spirituality of the self with the landscape," writes Dale Kerwin. The art historian Vivien Johnson has argued that these paintings have enough likeness to European cartography that they should be considered legal documents. "Like western topographic maps, these paintings are large-scale maps of land areas, based on ground surveys, with great attention to accuracy in terms of the positional relationships among the items mapped," she argues. "They can be used for site location, and because of their precision have the validity of legal documents—they are Western Desert graphic equivalents of European deeds of title."

Some people disagree. Australian anthropologist Peter Sutton has argued that the paintings can't be used for site location because someone unfamiliar with the country wouldn't be able to find their way using them. But the same might be said of a Google map: if you don't know what a car is and have never been on a road or been educated in the various symbols of modern cartography, a Google map is equally useless for site location. Both types of maps are dependent on a body of knowledge brought to them by the traveler; what is esoteric depends on the viewer. "European maps are not autonomous," writes the academic David Turnbull in his book *Maps Are Territories*. "They can only be read through the myths that Europeans tell about their relationship to the land." Turnbull has argued that maps themselves are metaphors for the cultures that created them—unless they are drawn on a foot-to-foot scale, their accuracy and reality has to be considered to be a point of view rather than a neutral or empirical depiction.

On October 19, 1972, David Lewis put his navigation research aside to undertake the first solo circumnavigation of Antarctica in a sailboat. It was a grueling trip that Lewis barely survived; he capsized three times and ultimately abandoned his boat, the *Ice Bird*, at Cape Town on March 20, 1974. When he arrived back in Australia, Lewis visited the desert again to undertake another trip with Jeffrey Tjangala and Yapa Yapa Tjangala. Their journey would start in Yayayi, a newly created Aboriginal community, and end in Jupiter Well some 370 miles to the west. Along for the ride would be Fred Myers, an American PhD student who had been living at Yayayi since June 1973, conducting field research for his doctorate in anthropology. In documentary footage from that time taken at the settlement, Myers, with brown hair and glasses, a cigarette in one hand and a reporter's notebook in the other, can often be seen in the background, quietly observing the gatherings and daily life of the Pintupi. He was particularly interested in documenting the creation of paintings at Yayayi, which were being purchased and then sold in Alice Springs. Myers's connection to the Pintupi and the politics, culture, and art of the Western Desert has lasted over four decades. In his 1985 book *Pintupi Country, Pintupi Self: Sentiment, Place, and Politics among Western Desert Aborigines*, Myers describes the scenery of Australia with deep affection:

> It is a stark country, known to Europeans as an arid and dangerous place, but its red sand, flat scrubby plains covered with a sparse pale greenery, and craggy, long-eroded hills lie in muted beauty beneath an awesomely blue sky. One cannot escape its immensity and its calm. The paleness of its colors seems always to be a kind of ghostly habita-

tion of color, barely corporeal. . . . In the enduringness
of this landscape, Aborigines see a model of the continuity
they aim to attain in social life, a structure more abiding
and real than their transitory movements on its surface.

———

On a dismally cold February morning I walked to Washington
Square Park in lower Manhattan and took an elevator up a gray
high-rise to Myers's office at New York University. Inside, my
spirits were lifted by the clutter of objects and books Myers has
collected over more than forty years of fieldwork in Australia. He
began opening file drawers and pulling out topographic maps of
the area surrounding Papunya, showing me the route he had taken
while traveling with Lewis and the Tjangalas. When the trip
started, Myers was already aware of the Aboriginal capacity for
uncanny navigation skills; even young children seemed incapable of losing their bearings. "Friends of mine have prodigious
memories," he told me. Often they were bewildered by the fact that
Myers could get lost so easily. "I can tell you kids have it by the
time they are seven or eight," he said. Sometimes when he needed
directions while driving, his companions would say in disbelief,
"You've been there, you saw it before! The road goes this way. Follow the one to the north." Myers laughed. "I would be driving in
this mulga scrub and worried I'm going to get a flat tire or rip
the transmission out and they're saying, 'North! North!' What
the fuck? I don't know which way north is."

Myers scrolled through files on his computer, looking for digital reproductions of photographs he had taken on the trip with
Lewis in 1974. In one, a twentysomething Jeffrey Tjangala stands
in front of the extinguished ashes of the previous night's fire holding a white enamel cup in one hand; it is morning and the camp is
being broken down before they move out for the day. Jeffrey
wears a peach plaid shirt, his dark jeans held up by a worn brown
leather belt and silver buckle; he's wrapped a piece of cloth as a
headband around his black hair. Behind him, Yapa Yapa Tjangala

stands at a slight angle from the camera, wearing a denim jacket and broad felt hat that casts a shadow over his eyes. They could have been members of the Jimi Hendrix Experience. "Jeffrey James and Yapa Yapa were two of my closest friends," Myers told me. "Jeffrey died a few years ago, an incredible person who more or less single-handedly got his people back to their country on the Canning."

Myers told me that he believes that Aboriginal contemporary art pieces are more like conceptual maps than geographical ones. "The placement of landscape features and their arrangement is rarely if ever replicated with actual geographical orientations," he said. "They are more like mnemonics of places. Some paintings have more obvious useful information than others." For individuals, he said, "I would say that most of their knowledge is very specific, of having walked it before. They can do other things but mainly they are remembering walking with their parents." Dreaming stories themselves have multidimensional purposes, Myers continued. "It's one form in which knowledge and direction and ecology are encoded. People also get rights through these stories, it links people, and it does provide them with an understanding of the landscape in these faraway places. It's a skeleton on which a lot of geographical knowledge can be placed. How do you remember this stuff? Well, my father grew up in this place, that's where these ants were. It's a way of condensing knowledge." The intricacy of this knowledge is remarkable. Even today, after all the decades he has spent in Australia and despite his fluency in several Aboriginal dialects, Myers said he still struggles to understand the language through which Aboriginal people give directions to one another.

After Lewis arrived in Yayayi, the party headed west for their seven-day trip. For Myers, one of the trip's most memorable moments was when something truly unusual happened: the Tjangalas lost their bearings. Lewis was also surprised by the episode and would write about it in several journal articles as well as his memoir *Shapes on the Wind*. The group had decided to go to a place called Tjulyurnya, a Dreaming site where dingos had driven two lizard men underground and left behind a pattern of triangular yellow stones. The Tjangalas wanted to get some *mulyarti* wood for spears and bring some of the sacred stones back to Yayayi. Tjulyurnya was twenty-six miles from their camp. Lewis carefully recorded their route to the Dreaming site across a terrain covered in spinifex, low sandhills, and "undulations hardly deserving the name of hills."

1. Seven kilometres a little south of west to Namurunya Soak, a tiny hollow that seemed, to my eyes, to have no identifying marks at all.
2. Thirteen kilometres south-west to a site where kante, sharp flints used for stone knives, were found. This was beside a low rise in the ground.
3. Five kilometres south-east, then round the end of a sandhill to Rungkaratjunku, a sacred site.
4. Winding in and out between low sandhills, generally west-south-west, sixteen kilometres to Tjulyurnya rock hole by a little hill. The sacred place was two kilometres further on.

When they arrived at Tjulyurnya, they gathered some of the stones and planned to make camp for the night. But the Aboriginal men started to worry. Had they made a mistake bringing white people to the Dreaming site? Perhaps they shouldn't have moved any of the stones. Finally, they decided it was best to get back to their previous camp as fast as possible. Everyone got into the two vehicles and shut the windows tightly in order to keep out devil dingo sprits. That's when disaster seemed to strike.

"It got so dark we lost the track," said Myers. (Lewis speculated that the headlights also effaced their night vision and masked the terrain.) "And we drove until we came right around back to where we had started. The men were freaking out because they thought the spirits were dragging us back. The same thing happened again when Jeffrey drove and now they were really freaking out." Finally their non-Aboriginal mechanic, David Bond, navigated by keeping the Southern Cross in the righthand window to take an easterly course, getting them back to their original camp. What Lewis surmised was that despite the Pintupi's faultless navigation skills, they were incapable of wayfinding according to the stars. But in Myers's opinion the problem was not the stars but the speed of the cars they were driving. "When you are walking, you don't lose your sense of direction. They know where the stars are, they know where the star comes up but I think they don't need it," he told me. "How they process [directionality] as they are walking is with bodily orientation. They don't stop to say, which way is the north? It's not a matter of calculation, or a simple cognitive thing. It's very much a constant tracking."

By the time Lewis's research in Australia ended, he had come to believe that his companions' wayfinding abilities came down to "some kind of *dynamic image* or *mental 'map', which was continually updated* in terms of time, distance and bearing, and more radically *realigned at each change of direction*, so that the hunters remained *at all times* aware of the precise direction of their *base and/or objective*." Unbeknownst to him, at the same time he was roaming the desert with the men from Papunya, two neuroscientists over nine thousand miles away in London were developing a very similar theory of human navigation according to a mental map in the brain.

SPACE AND TIME IN THE BRAIN

In the early 1970s a young American scientist by the name of John O'Keefe went in search of one thing but got lost and discovered another. Like so many scientific breakthroughs, curiosity, skill, and luck colluded, and in this case, the accident garnered him a Nobel Prize. O'Keefe was interested in recording single neuron activity in the amygdala—the place of emotional learning. One day at his laboratory at University College London he attempted to implant a microelectrode in a rat's somatosensory thalamus, the place where sensory perceptions are processed. However, he used the wrong coordinates and ended up inserting the electrode into the rat's hippocampus. As the single cell O'Keefe was recording began to fire, its pattern struck him as peculiar. The cell's activity seemed to be strongly correlated with the animal's locomotion. His interest piqued, O'Keefe abandoned his research on the amygdala and began recording single hippocampal cells of rats while they were eating, grooming, and exploring.

He wasn't the first to record these cells: a Russian neuroscientist named Olga Vinogradova had recorded them in rabbits in 1970 and thought they might be responses to stimuli. O'Keefe inferred a different significance. As he wrote, "Over a period of months, I began to suspect that their activity didn't depend so much on what the animal was doing or why it was doing it but had something to do with where it was doing it. Then on one electrifying day I realized with a flash of insight that the cells were responding to the animal's location or place in the environment."

O'Keefe began changing aspects of the rat's environment and watched the effect on hippocampal cell activity. Even when he turned the lights off in a familiar maze, the cells continued to fire. It didn't matter which direction the rat was facing, or whether rewards were taken away or changed. The only stimulus that seemed to matter to the cells was the rat's location. Rather than respond to the changes in stimuli, the cells were signaling an abstract concept of space. O'Keefe called them place cells.

In the decades since, these cells have fascinated researchers with their plastic, almost magical properties. The pattern of their firing corresponds to a location in the environment, and so, amazingly, an animal's location can actually be reconstructed just by the firing rate of place cells. This means that scientists can track the neural activity of the rat and, based on this information alone, accurately infer where the rat is in physical space in real time. Studies have shown that once a place cell encodes space, a process that seems to happen within a couple of minutes of a novel experience, it can retain the same firing pattern for months, indicating a role in spatial memory. Place cells have been recorded when a rat is sleeping and seem to fire in similar patterns to the rat's prior experience, and it has been hypothesized that sleep might be involved in consolidating memories of the spaces the rat has recently explored. These cells are also capable of remapping themselves, meaning different patterns of the same cells will fire when a rat is placed in a different environment.

Almost immediately after discovering place cells, O'Keefe was struck by the idea that they seemed to prove a little-known theory that had been proposed over thirty years earlier, long before advances in technology allowed scientists to record individual neuron activity. "In thinking about these results over the next day," wrote O'Keefe about his discovery, "I was assailed by a montage of ideas about the potential significance of this finding: the first was that it might mean that the hippocampus was the neural site of [Edward] Tolman's cognitive map, a vague hypothetical construct that he had used to explain some aspects of rodent maze behavior but which had never gained much acceptance in the animal learning field and which was little discussed in the 1960s." O'Keefe experienced a "prolonged euphoria of the classical Archimedean type": maybe he had found the cognitive map.

The word *labyrinth* comes from the Greek *labyris,* meaning "double ax," a symbol for the Minoan goddess of Crete. It was King Minos who asked Daedalus to design a labyrinth so complex it could imprison the Minotaur, which was later killed by Theseus, who followed Ariadne's thread to find his way out. The word *maze* likely had an original meaning of "to be lost in thought," and in Middle English it meant to confuse, puzzle, or dream. Using rats in mazes to glean insights into behavior and spatial cognition is a tradition over a century old. In the 1890s a young psychology student in Chicago observed how the rats at his father's farm made runways under the porch of an old cabin to their nests; when the runways were revealed, they looked just like a labyrinth. Maybe psychologists could use labyrinths to find out more about what they called the rats' home-finding abilities and test their memory and learning?

One of the student's colleagues was the experimental psychologist Willard Small, who, inspired by these conversations, became the first person to design a maze for rats. For inspiration he used the famous hedgerow labyrinth, a trapezoidal maze full of twists and dead ends, created at London's Hampton Court in the late seventeenth century. He put wire mesh on a six- by eight-foot platform and gave it six culs-de-sac. Then he meticulously detailed each and every step the rats took to explore it and marveled when several blind rats found their way as easily as the others.

Experiments like Small's became increasingly popular, so much so that in 1937 Edward Tolman addressed a conference of his colleagues and said, "Everything important in psychology . . . can be investigated in essence through the continued experimental and theoretical analysis of the determinants of rat behavior at a choice-point in a maze."

The typical experiment in Tolman's time was to withhold food from the animal and put the rat at the entrance of a labyrinth that had several blind turns and a food box at the end of the correct path. The researchers would time how long it took the animal to find the food and then test the rat again and again every twenty-four hours. Eventually, every rat would learn where the blind turns were and would follow the most direct route through the maze straight to the food. But sometimes the rats in these experiments would do things the psychologists couldn't explain. In 1929, one scientist reported that his rat learned a maze and then instead of running through it again to get a reward, pushed off the cover of its starting box and ran across its top, beelining for the food and bypassing the whole experiment. This behavior prompted similar questions to those asked by Felix Santschi, who studied Tunisia's desert ants. How could rats infer spatial relationships that allowed them to take shortcuts? The popular scientific explanation was that all animal behaviors, including those demonstrated by rats in mazes, were the result of stimulus-response. Rats see, smell, and hear stimuli from the environment and process these through sense organs, which transmit signals to muscles. Learning to turn left or right through a maze was the result of this behavioral conditioning.

Tolman, a graduate of the Massachusetts Institute of Technology, was one of the first psychologists to doubt that theory. He called its adherents the "telephone switchboard school" for their mechanistic reductionism. Tolman thought that rats had brains capable of learning routes and building representations of the environment. Rather than thinking of them as mechanized automatons with inputs and outputs, Tolman described the rat's mind as containing a "cognitive-like map of the environment." He specified that this cognitive map wasn't just a strip map of the particular paths that led to the food but a comprehensive map that included the location of food and the surrounding space, enabling rats to find novel routes. The idea of a cognitive representation of space was a radically different explanation for rat navigation. And Tolman went so far as to argue that the same mechanism was at work in humans; his classic paper on the subject, published in 1948 in *Psychological Review,* was called "Cognitive Maps in Rats and Men."

At the end of the paper Tolman elucidated an argument that he called "brief, cavalier, and dogmatic." Was it possible, he wrote, that many people's social maladjustments could be interpreted as the result of having too narrow and limited cognitive maps? In one example, Tolman wrote about the tendency for individuals to focus their aggression on outside groups. Poor southern whites, he wrote, displace their frustration with landlords, the economy, and northerners onto black Americans. Americans as a whole displace their aggression onto Russians and vice versa. He wrote,

> My only answer is to preach again the virtues of reason— of, that is, broad cognitive maps. . . . Only then can these children learn to look before and after, learn to see that there are often round-about and safer paths to their quite proper goals—learn that is, to realize that the well-beings of White and of Negro, of Catholic and of Protestant, of Christian and of Jew, of American and of Russian (and even of males and females) are mutually interdependent. We dare not let ourselves or others become so over-emotional, so hungry, so ill-clad, so over-motivated that only narrow strip-maps will be developed.

For decades after Tolman first wrote about it, the cognitive map remained an obscure concept, and few psychologists, let alone

animal behaviorists, were interested in it. Tolman himself probably never guessed that such maps might have a neural basis, the
product of a cognitive mapping *system* located in a specific place in
the brain. Sadly, he passed away in 1959, long before O'Keefe began
recording place cells in the rat hippocampus.

———

Before moving to London, O'Keefe had studied at McGill University's psychology department in Montreal, a mecca at the time
for physiological psychology, and became friends with another
graduate student, Lynn Nadel. Both were New Yorkers—O'Keefe
was born in Harlem, and Nadel was born in Queens—and both
had studied with the psychologist Don Hebb, who encouraged his
students to theorize about the neural basis of cognition and test
their ideas. O'Keefe and Nadel stayed friends even after leaving
Montreal. Nadel went to Prague to do a postdoctoral fellowship,
and when the Soviet Union invaded Czechoslovakia in 1968, he
drove his wife and kids across Europe to O'Keefe's home in London. Nadel shared O'Keefe's interest in the hippocampus, and he
joined University College London to collaborate with O'Keefe on
his research into the cognitive map.

Initially they set out to write a single paper proposing the hippocampus as the source of Tolman's cognitive map. The paper grew
to hundreds of pages long. Along the way they realized that in order to refute the standard theory of animal learning as stimulus-
response, they would have to master the theory first. Eventually
they sent their research to fifty different colleagues to solicit feed-
back, and six years later, instead of a paper, they had a book, one
that would end up influencing the trajectory of the next forty years
of neuroscience. Called *The Hippocampus as a Cognitive Map*, it
was published in 1978 and dedicated to both Tolman, "who first
dreamed of cognitive maps in rats and men," and Hebb, "who
taught us to look for those maps in the brain."

———

O'Keefe and Nadel's tome begins with a most basic assertion—that space is one of the most important forces shaping the human mind.

> Space plays a role in all our behavior. We live in it, move through it, explore it, defend it. We find it easy enough to point to bits of it: the room, the mantle of the heavens, the gap between two fingers, the place left behind when the piano finally gets moved. Yet, beyond this ostensive identification we find it extraordinarily difficult to come to grips with space. . . . Is space simply a container, or receptacle, for the objects of the sensible world? Could these objects exist without space? Conversely, could space exist without objects? Is there really a void between two objects, or would closer inspection reveal tiny particles of air or other matter? . . . Is space a feature of the physical universe, or is it a convenient figment of our mind? If the latter, how did it get there? Do we construct it from spaceless sensations or are we born with it? Of what use is it?

They believed that the purpose of the hippocampus was to process and construct models of the physical universe in which we exist. This was a controversial claim. The field of cognitive neuroscience treated many learning processes as diffused throughout several interconnected systems. Now Nadel and O'Keefe were claiming that when it came to the spatial mapping system, the physiology of a single circuit deep in the medial temporal lobe had evolved specifically to create and store spatial representations. But why did the "Constructor of Brains," as they jokingly referred to the mind's creator, require that spatial mapping occur in one specialized part of the brain?

The reason, they argued, was that *space itself is special.* Whereas color, motion, and other properties of objects in the physical universe can be taken away, so to speak, space has a unique status: it is an "ineliminable property of our experience of the world." The first fifty pages of *The Hippocampus as a Cognitive Map* is a survey of theories of space throughout Western philosophy. The authors jump from Isaac Newton to Gottfried Leibniz to George Berkeley to Immanuel Kant, pointing out that these philosophers and many other physicists and mathematicians fall into one of two camps: either they conceived of physical space in absolute terms or in relative terms. An absolute view of space, espoused by Newton in the seventeenth century, is that it is a fixed framework—a container, so to speak—in which objects exist. In contrast, the relative view of space is that it is made up of the relationships *among* objects and can't exist independently of those relationships. Berkeley, Leibniz, and David Hume went so far as to argue that our minds can't even access the physical world—because they questioned the physical world's very existence. To them, space was manufactured by the human mind. Over his lifetime, Kant swung between these arguments before eventually publishing the *Critique of Pure Reason* in 1787 in which he argued that space was absolute, but only because the mind was innately equipped to organize it that way. Or as O'Keefe and Nadel put it, "Space was a *way* of perceiving, not a thing to be perceived." For the two neuroscientists working centuries later, Kant's ideas were not only inspiring; they felt that they had discovered the neural basis for his philosophical model of an a priori spatial faculty.

In their book, O'Keefe and Nadel argued that both an absolute and relative understanding of space were important for humans. The Constructor of Brains had "hedged his bets and incorporated both systems into his invention." An organism experiences space in relationship to itself (egocentric), but the brain is also capable of what they called "nonegocentric cognition," or allocentric perspective, an ability to objectively represent the environment in three-dimensional space. This is the cognitive map in the hippocampus.

Nadel and O'Keefe based their theory on several hundred studies from "lesion literature," in which animals and in some cases humans with damage to their hippocampi were tested on tasks to try and understand what aspect of cognition was affected. When researchers took away the part of the brain responsible for organizing space, the consequences were devastating. For example, in 1975 O'Keefe and Nadel and several of their students took thirty-two male rats and gave sixteen of them lesions by cutting open their skulls and crushing their fornix—the nerve fibers that act as an output from the hippocampus—with jeweler's forceps. All of the rats were then kept from water until they were thirsty, and tested repeatedly on how quickly they could find water in a room. The location of the water never changed, but those rats that had the lesions were incapable of remembering where it was or the route to get there, and thus of doing what is called place-learning. The animals searched for water each time as though it was the first time they were taking the test; they had lost their cognitive mapping capability. These studies were the scaffolding for Nadel and O'Keefe's hypothesis regarding the key function of the hippocampus. The hippocampus's cells, they argued, encode space in an allocentric (nonegocentric) framework, the map. Then the animal uses that map for navigation, computing the distance and direction between landmarks, orienting itself and inferring spatial relationships.

Before O'Keefe discovered place cells in the early 1970s, neuroscientists knew that the hippocampus was involved in memory, though what *kind* of memory was disputed. One of Nadel and O'Keefe's professors at McGill was the neuropsychologist Brenda Milner, the first person to have studied the patient H.M. and written a paper about the nature of his amnesia after the removal of part of his temporal lobe to treat severe epilepsy. Milner recognized that there were different types of memory systems and learning, and that H.M.'s amnesia was episodic in nature. But later theories of the hippocampus and memory argued that the hippocampus was responsible for something else as well: semantic memory, the recollection of facts. Reconciling Milner and these other theories of the hippocampus was one of the major challenges facing O'Keefe and Nadel. Their theory predicted that the hippocampus was the core of a neural system that provided a spatial framework for storing memories of what happened in a particular place, not facts but events. "The idea was that episodes were built upon the basic spatial framework through the addition of a linear sense of time amongst other higher-order cognitive capacities."

It wasn't until the 1990s and the advent of virtual reality—the ability to create computer simulations of large-scale environments—that neuroscientists could use MRI on immobile people to fully understand what parts of the brain were activated during navigation and memory recall—thereby confirming this idea. The earliest virtual reality tests used a first-person shooter game, *Duke Nukem*, stripped of all of the guns and fighting, leaving only the mazelike environment that subjects could wayfind through. In 2001, O'Keefe and several other researchers at University College London designed a study whereby epileptics who had undergone lobectomies in their right or left temporal lobes were asked to explore a town in the video game environment for about an hour, during which they encountered different characters. They were then tested on their ability to draw maps of the environment and their memory of the events. They found that whereas those with right temporal lobe lobectomies were impaired in navigation and spatial memory, those with left-side lobectomies were impaired in the episodic memory tests, suggesting that the hippocampus was indeed a critical part of the brain for both cognitive mapping and episodic memory.

In subsequent years, scientists discovered other critical cells in the hippocampus and a remarkable level of plasticity in hippocampal physiology. Some of these other cells include head-direction cells, which discharge in relation to which way our head is pointed on the horizontal plane, and grid cells, which fire as we roam an environment and build a coordinate system for navigating. There's also evidence that the richness and complexity of an environment influences the quantity of neurons in the hippocampus. In 1997, for instance, three researchers, including Rusty Gage at the Salk Institute, found that mice exploring enriched environments—paper tubes, nesting material, running wheels, and rearrangeable plastic tubes—had forty thousand more neurons than a control group. These additional neurons resulted in an increase in hippocampal volume of 15 percent in the mice and significant improvements in their performance on spatial learning tests. The researchers concluded that a combination of increased numbers of neurons, synapses, vasculature, and dendrites led to the animals' enhanced performance on the tests.

Today, an even fuller picture of how hippocampal cells interact and build spatial representations for orientation and navigation has emerged. As Kate Jeffery and Elizabeth Marozzi explained in *Current Biology*, multiple sensory systems, from vision to touch and olfaction, converge upstream of the hippocampus and are "combined into supra modal representations such as landmarks, compass cues, boundaries, linear speed," which are then passed on to the place cells. At the same time, head-direction cells give us a sense of direction, firing only when the head is pointed in a particular direction, like a kind of neural compass. Border cells seem to signal the distance and direction from a boundary that could be an obstruction, gap, or step. Grid cells are thought to represent space at different scales by using both environmental and self-motion information to generate information about distance. They fire in an environment in a fascinating pattern: a hexagonal array that extends in all directions, and they are one synapse upstream of place cells. While the interaction among these different cells is still somewhat mysterious and the topic of much research, it's likely that grid cells send information to the place cells used for path integration while also receiving information in return. Very accurate navigators seem to show more hippocampal activation and engagement, and navigational experience itself seems to increase volume of the brain, as the study of London's taxi drivers originally showed. The cognitive mapping system is, surprisingly, not dependent on vision. There is evidence that blind people formulate cognitive maps. Blind individuals use kinesthetic and motor signals to dead reckon, and seem to be better at this than sighted individuals.

Matt Wilson is a neuroscientist at MIT. He calls the classic experiments of running rats through mazes and listening to their brain cells "eavesdropping," and he has been doing it for years to try and understand what this system of cells has to do with memory. Testing the relationship takes ingenuity. "If you damage the hippocampus in humans or rodents, you will lose the capacity to form memories of life experience. Now, it's hard to ask those life-experience questions to a rat. But what you can do is test them on another kind of memory: just ask a rat to go back to a place it's already been. Rats have very good spatial memory." The connection between spatial navigation and memory of life experiences, according to Wilson, is time. "Both [navigation and memory] depend on a critical function, linking things in time," he said. "It is how you put the pieces together, how you create an internal narrative of your experience. It's not a record, or videotape of experience. It involves evaluating, selecting, and sorting things. Rats create an experience of moving around in space. We create the stories of our lives."

———

How can we be so sure that space was the primary concern of place cells? What if space just matters *more* to rats, the favored study subject of tens of thousands of maze experiments conducted since the early twentieth century, and there are other domains of experience the hippocampus is sensitive to? Some neuroscientists believe hippocampal cells are actually implicated in a far greater scope of human cognition than spatial representations, and they doubt that our brains are even really building representations that are structurally analogous to an allocentric map. Maybe the cognitive map is much more flexible than that, and the hippocampus encodes and builds maps for *many* dimensions of human experience beyond space—everything from time to social relationships, sound frequencies, even music.

One warm fall day, I walked from the MIT campus, where Edward Tolman had once studied and the amnesiac patient H.M. had spent countless hours being observed, and crossed the Charles River to Boston University to meet one of the most prominent dissenters of the cognitive map theory—Howard Eichenbaum, the director of the Center for Memory and Brain and the Cognitive Neurobiology Laboratory. I climbed the stairs to Eichenbaum's second-floor office and knocked on the door. I was greeted by a white-haired, mustachioed man from behind a desk covered in stacks of papers, likely related to his work as editor in chief of the scientific journal *Hippocampus*. Hanging on the wall in a frame behind him, I saw a poem, "The Experiment with a Rat" by Carl Rakosi:

> *Every time I nudge that spring*
> *a bell rings*
> *and a man walks out of a cage*
> *assiduous and sharp*
> *like one of us*
> *and brings me cheese.*
> *how did he fall*
> *into my power?*

He put his feet up on a chair and asked me, "So what is navigation anyway?"

I laughed. It was the simplest question. But despite thinking about almost nothing else for several years, I had yet to find a simple answer. It is fundamentally the task of getting from one place to another. But it can entail so many different strategies in both animals and humans, so many scales and perspectives, that it hardly seems to fit under one action, process, or skill. Instead, it involves multiple arrays of cognition and possible problem-solving techniques. Thus far, scientists have created a multitude of categories to try and capture it. Vector navigation involves moving along a constant bearing relative to a cue that could be magnetic, celestial, or environmental. Piloting is defined as navigating relative to familiar landmarks. True navigation generally means wayfinding toward a distant, unseen goal. Dead reckoning, also called path integration, is keeping track of every stage of a journey in order to compute one's location.

As it turns out, both rats and humans are the worst at path integration, which is precisely the kind of navigation that cognitive-map theorists propose the hippocampus does. In Eichenbaum's opinion, this is extremely problematic. "One of my complaints about the path integration theory is how *bad* we are at it," he said. Dead reckoning is applicable in short distances at local scales, but it is a strategy that isn't actually advisable in real-world navigation because it is so prone to accumulated error (except, it would seem, for those who have mastered complex environments like the Arctic tundra or Australian desert). Can the navigational capacities of humans be fully explained by the cognitive map theory of the hippocampus, or is there more going on?

Eichenbaum most readily describes navigation as what it is not. "I think navigation is not about Cartesian maps," he offered. "It's a story or memory problem." The hippocampus is not so much about spatial memory, he said, as it is about "memory space." Parsing this distinction is important. True navigation, in his opinion, is what happens when we travel to an unseen place. It requires planning a future (envisioning the place we want to go), calculating or remembering the route to get there (a sequence or narrative), and then orienting to ensure we are on the right track, often by comparing our memory (or perhaps a description we've been told) to our real-time perception of movement through space. "There are huge memory demands to solving the problem of navigation," he said. "Memory steps in at every moment."

Space and its role in hippocampal function has been oversold, says Eichenbaum, for whom space is just one of many "fabrics" that we hold memory in. He believes that the hippocampal cells called place cells are much more flexible and capable of adapting to different dimensions. One of those dimensions is temporal, and for this reason, Eichenbaum doesn't call them place cells, he calls them time cells. "Time is a philosophically interesting question. Do we make it up?" he mused. "As you navigate, you are moving in space and time together, and the hippocampus is mapping both." His research has led him to believe that the organization of our episodic memories is supported by these time cells, and that mapping sequences of memories in time is just as critical to navigation as mapping geographic space. The trick is trying to design experiments that can demonstrate the difference because "you can't usually parse space and time."

He motioned me to his desk and opened a video file on the computer. I was looking down at the faint outline of a healthy, plump white rat with black markings. Its head was obscured by wires attached to the electrodes inserted in its brain. Eichenbaum had conducted this particular experiment in the laboratory across the hall a few years earlier. It appeared to be like so many others. A rat is released in a figure-eight maze with a reward at the end. But this one had a treadmill at the stem of the maze. Before the rat could find its way to the reward it had to step on the treadmill, which was programmed to randomly speed up and slow down. As the rat started running in place at these different speeds, the electrodes in its brain recorded the firing of three different hippocampal cells, represented by a colored pixel on the screen. "Watch closely," said Eichenbaum. "First you'll see a blue dot, then a green, and last a pink dot."

As the rat started to run, I saw each cell fire in the order Eichenbaum described. I could tell that watching the video still thrilled him four years later. But what did these colored pixel-neurons prove? By holding behavior (running) and location (in place) constant, and randomizing the treadmill speed, Eichenbaum had decoupled the distance the rat traveled from the time it spent running and could track which neurons were mapping each variable. The results show that the hippocampus was encoding *both* time and distance simultaneously. Then, when the treadmill stopped and the rat continued through the maze, the very same neurons began to fire because they were encoding *space*. Experiments like these, in which hippocampal cells "map" multiple dimensions, are why Eichenbaum believes the hippocampus is capable not only of organizing physical space but of creating "temporally structured experiences into representations of moments in time."

After years of studying rats in mazes, Eichenbaum has come to understand the hippocampus as the "grand organizer" of the brain. "It's organizing and integrating all these bits and pieces of information in a contextual framework," he said. "It does create a map. I'm all for the cognitive map in the original sense that it's a map where you put stuff to remember where they are in relationship to each other. That is a specific, limited, concrete sense of moving in geographic space and how did I get from here to there. The other sense is this abstract term, how did I navigate graduate school? What's the path to the presidency? In human language, these are both legitimate. But which one is the hippocampus? Is it the specific one or the generic one? I think the hippocampus could mean to map things in time. And there are other spaces in addition to geometric space. It doesn't have to be Euclidean or linear. That's just a really good example of what the hippocampus does, but it has other functions."

In the last five years there has been more interest in designing tests to explore what those other spaces could be. A few years ago, a team of researchers in New York and Israel wondered whether the hippocampus could map social space: the relationship and interactions among individuals with different roles and levels of power. They asked individuals to participate in a role-playing game in which they moved to a new town and had to find a place to live and work, and they found that the hippocampus was activated during the tasks, indicating it's a circuit that is important for "navigating" social relationships. Another study, authored by Sundeep Teki and others in 2012 and called "Navigating the Auditory Scene," found that professional piano tuners had something in common with London's taxi drivers: more hippocampal gray matter. The more years a person spent tuning pianos, the larger this part of their brain was. Sound, in their case, was the space that the hippocampus mapped. Different pitches and beat rates were landmarks, and routes were created from one previously tuned note to another. A study in the same journal two years earlier reported that musical training actually induces plasticity in the hippocampus. After just two semesters of training, the researchers saw evidence in fMRIs that music academy students' hippocampi had enhanced responses to hearing sounds. Had their hippocampal cells become music cells?

Eichenbaum thinks that results like these are perhaps *more* faithful to the original idea of the cognitive map described by Tolman back in 1948. A close reading of that now-historic paper reveals that he thought the cognitive map might be multidimensional, a tool capable of mapping multitudes of life experiences. And these new studies are also relevant to how we answer the question Eichenbaum first posed to me: What is navigation really? Insights into time cells, social space, and music highlight how complex human navigation in the brain is: not just a calculation based on reading a Cartesian map but an unfolding memory or a narrative sequence, human relationships, sensory experiences, personal history, or paths into the future. "The hippocampal system," Eichenbaum once wrote, is "encoding events as a relational mapping of objects and actions within spatial contexts, representing routes as episodes defined by sequences of places traversed."

Sometimes that sequential story is geographic. Sometimes that episodic narrative is who we met and the words that were spoken. Sometimes that memory is a piece of music that carried us on a journey.

The notion of a map guiding our movements is so deeply pervasive, a metaphor so cherished by the Western mind that it can feel
nearly impossible to transcend. How could we know our way
without a map? How can most of us, even children, sit down and
draw maplike representations of familiar places if we don't already
possess one inside our heads?

Throughout history, scientists have turned to material artifacts as metaphors for understanding the processes of life. Kepler
described the universe as a clock. Descartes described reflexes as
a push-and-pull system typical of sixteenth-century technology.
Tolman's contribution was to depart from the telephone switchboard metaphor to a map. Today it's common to see the human
brain likened to a computer, and the hippocampus to a GPS. Can
these metaphors really capture the complexity of biology, or do
we reach for them because we lack imagination for what's really
at work? "The cognitive map is a metaphor for what the brain
does," the neuroscientist Hugo Spiers told me, "but the problem
with maps is they are complicated ideas in and of themselves. They
are a type of metaphor already."

The philosopher William James called this problem the psychologist's fallacy. James was concerned that scientists so often
mistake the product of reflection and analysis of our experience
as characteristic of immediate experience. Yet when we reflect
and analyze, we are already stepping outside our direct experience
in order to give an account of it, already beginning the process of
reaching for metaphors that can only fall short in capturing our
experience. Often those metaphors and models we reach for are
influenced by human tools, not innate cognitive processes. James's
philosophy is called "radical empiricism," and he believed that
humans were capable of directly and objectively perceiving the
world.

If the map is a psychologist's fallacy for understanding wayfinding, what is a more accurate metaphor? Consider how you get
from your home to work. Do you see a picture of the whole route,

a bird's-eye view from above, and begin charting your course? Likely not. Rather, you know your starting point and the series of decisions you will make, and you have a visual memory of the route that follows. It's an experience that is perhaps more akin to recalling a melody, a fact that was pointed out to me by Harry Heft, a professor of psychology and environmental studies at Denison University in Ohio. "When I think about getting to work, it's like I want to start humming or singing a song. I don't hum the whole song before I begin. I think, how does it begin?" said Heft. "As with humming a melody, I might get lost at some point and then I would stop and keep thinking of the thread again— what happens after this point? I see the analogy of navigation and music as quite direct because they are both temporally structured information." Maybe the metaphor at the heart of navigation is not following a map but listening and intuiting the progress of a piece of music.

Heft's academic lineage traces back to William James. He studied with James Gibson, pioneer of environmental psychology, who was taught by psychologist E. B. Holt, who was James's student. Like his mentor Gibson, Heft does not believe cognitive maps are involved in wayfinding. Sure, we can conceptualize a maplike layout of our surroundings if asked to do so. But such Euclidean coordinate maps are not the foundation of our spatial knowledge, he told me. "They don't exist as we travel from one place to another as a picture in our head. Of course, we are capable of creating all *kinds* of images in our head. We can picture family members when they're not present. But when they are present, we perceive them directly. We only produce images of people at a remove from immediate experience. A cognitive map is like that: it doesn't guide us in an ongoing manner. We can produce a map to try and get our orientation, but it's not the foundation of wayfinding."

Heft has written that the human ability to think in terms of configurational, Euclidean-cartographic fashion is a historical development stemming from the cultural invention of maps, such as Ptolemy's *Geographia,* and the European expansion of economic and political power in the fifteenth and sixteenth centuries. Today, the ubiquity of maps and our constant exposure to them has made it much easier to assume they describe a basic mental process. "When I read the animal literature and the insect literature and folks are talking about cognitive maps, I just scratch my head," Heft said. "From my perspective, that's James's notion of the psychologist's fallacy, of imposing the concepts onto the process you're trying to study. To talk about the hippocampus as having a GPS—that's crazy. It's imposing these concepts of a much higher order of nature to a level of functioning where it doesn't fit."

Heft was a graduate student in psychology in the mid-1970s when he discovered James Gibson's 1966 book *The Senses Considered as Perceptual Systems,* which described environmental psychology and the idea that humans could directly perceive the world. While many of Gibson's peers ignored or criticized his work for its radical departure from the academic canon, others felt they had finally found answers to profoundly important questions about human experience. Heft was one of them. "It was like a religious experience," he said of reading the book. "I felt it was absolutely right." Forty years later, Heft can still recall exact passages that inspired him. Reading Gibson's assertion that visual perception wasn't based on sensory inputs or stimuli assembled into a mental representation but immediate ecological information was "like hearing violins" for Heft. "The brain is relieved of all this work we kept assuming we had to do," he realized. In 1975 Heft was finished with his dissertation and wrote a letter to Gibson. Could he come to Cornell for a year to learn about ecological psychology? The professor agreed, and when Heft arrived in the fall, he discovered a handful of like-minded pilgrims all doing the same thing: gathering around the elderly professor in order to understand the tenets and implications of his theories.

One night Gibson's wife, Eleanor, a respected psychologist in her own right and a professor at Cornell, held a party at their house and asked Heft for a favor. Gibson, then in his early seventies, taught a class once a week at SUNY Binghamton in the early evening. Would Heft drive him? He immediately said yes. "Once a week I would spend a couple of hours in the car with him," Heft said. "I was pretty new to the topic so I would formulate my sophomoric questions the week before and then we would chat all the way there and all the way back. One of the things I was wondering is: How are we doing this? How are we finding our way from Ithaca to Binghamton? In my program they had been talking about cognitive maps all over the place." At the time Gibson was still working on his book *The Ecological Approach to Visual Perception,* in which he talked about wayfinding and how it consists of a sequence of transitions, the stretches of connected sequences in what we perceived over time, that connect "vistas." This explanation stuck with Heft, and he has shot a series of sixteen-millimeter films with his students to explore the relationship between ecological information perceived over time and wayfinding. In one they created a "transitions" movie consisting of just the transitions along a route presented at ten-second intervals. Another film consisted only of vistas along the same route presented at ten-second intervals. Then he asked study subjects to watch either the transitions or the vistas film, or the entire unedited film three times, before transporting them to the start of the route. He found that those who had seen the transitions were able to navigate with greater accuracy, strengthening his belief in Gibson's view that transitions are critical for route-learning.

Heft now thinks that cartographic maps have influenced human thought so heavily they have obscured how much of our navigation is about the pickup of visual information over time as we travel through the environment. "Wayfinding to a specific destination involves traveling along a particular route so as to generate or re-create the temporally structured flow of information that uniquely specifies that path to the destination," he has written.

"This temporal approach requires a departure from standard ways of thinking about navigation—a shift made easier if instead of drawing a parallel between navigational knowledge and perceiving a pictorial map, we recognize that a more appropriate parallel may be between perceiving route structure and perceiving musical structure."

AMONG THE LIGHTNING PEOPLE

Maybe navigation is more like singing a song than following a map. If so, Aboriginal navigation according to songlines would seem to epitomize this strategy. I flew to Darwin in Australia's Northern Territory to meet an Aboriginal man by the name of Bill Yidumduma Harney, who had coauthored a paper in 2014 called "Songlines and Navigation in Wardaman and Other Australian Aboriginal Cultures." Harney had described how as a child in the Northern Territory he had been taught how to derive compass directions as well as use the stars as mnemonic devices to remember Dreaming stories. The Wardaman, whose ancestral homelands stretched across a 5,000-square-mile region, often traveled at night, when it was believed distances shrank. In order to find their way they memorized the stories associated with constellations, and to keep time they followed the movement of stars they associated with the Dreaming of crocodiles, catfish, and eagle-hawks. "We talk about emus and kangaroos, the whole and the stars, the turkey and the willy wagtail, the whole lot, everything up in the star we named them all with Aboriginal names."

Before I left for Australia, I had called Harney's coauthor of the paper, Ray Norris, on the phone. An astrophysicist with the governmental space agency Commonwealth Scientific and Industrial Research Organisation, Norris works to detect radio signals emitted from galaxies millions of light-years away in order to understand the evolution of the universe. But in his spare time, Norris contributes to a little-known field of study called archaeo-astronomy or ethnoastronomy: how ancient and contemporary indigenous cultures understand the celestial sky. Norris became interested in the field after getting his theoretical physics degree from Cambridge University in the 1970s; he studied Stonehenge and ended up surveying most of the stone circles in the British Isles. I wanted to talk to Norris because he was one of the few people I was able to find who had published any articles specifically about Aboriginal navigation in the last forty years. In fact, Norris is among just a handful of people, including David Lewis, who in the last *century* has written about how members of Australia's hundreds of language groups wayfind.

My first question to Norris was, why can't I find more on this subject? Norris told me there were instances while scouring the anthropological literature where he could see how close some academics had come to grasping the importance of songlines for navigation but for whatever reason hadn't considered it very interesting. They may also have been deterred; most Dreaming stories and places are considered so sacred and unknowable to the uninitiated that sharing them with outsiders could lead to extreme consequences (in some cases, immediate death).

One of the most vivid descriptions I found of the ritual inculcation of young Aboriginal people to the Dreaming stories and songs was in a book written by a white Australian. The elder Harney spent most of his life in the company of Aboriginal people, first as a cattle drover in the outback, then as an employee of the Australian government's Native Affairs Branch, and finally as the first ranger of Uluru—the sacred and iconic sandstone rock formation in central Australia. He wrote many books, including one with the anthropologist A. P. Elkin, documenting song-cycles. In *Life among the Aborigines,* he described the disrup-

tion to the culture and transmission of Dreaming songs in Arn-
hem Land.

> Dead too are the "song-cycles" that took weeks to record;
> epic chants taught by elders to the male initiates at night by
> taking them to the sacred places and chanting the lines
> over and over into the youth's body and head until he be-
> came saturated with the theme and, forgetting everything
> else, he repeated the chanting of the elders. When this oc-
> curred the elders would pause to listen, and if he was out in
> one word or syllable they would chant once more until he
> was word-perfect. So each line was learnt, and after years of
> teaching, the one with the best memory would become the
> "Song-man" to carry on the traditions of the tribe.

His name was Bill Harney, and he was the father of Yidumduma.

Now in his eighties, Harney's son had helped Norris to understand how Dreaming tracks not only mark routes on the ground
and describe the location of waterholes, landmarks, boundaries,
mountains, and lakes but also follow the movement of the stars
in the sky. For example, an Eagle-Hawk songline of the Euahlayi
people, who live in New South Wales and southern Queensland,
extends some fifteen hundred miles from Alice Springs to Byron
Bay and follows the stars Achernar, Canopus, and Sirius. Another
Euahlayi songline, the Black Snake/Bogong Moth songline, follows
the Milky Way to connect the Gulf of Carpentaria to the Snowy
Mountains, a distance of seventeen hundred miles.

Harney's descriptions of how he accumulated this knowledge
of the stars and their songlines as a child were lovely: in the bush
he spent his nights lying on the ground with other children and
elders, who translated the stories the stars were telling them. "If
you lay on your back in the middle of the night you can see the
stars all blinking. They're all talking."

On the phone Norris said, "To be an elder is not an honorary title. You really have to work on it. From puberty to forty, he would have been memorizing stuff all the time. He says he knows the name of all stars down to the naked eye."

"Really?" I exclaimed. "How many can you name?"

"As an astrophysicist, my knowledge of the stars is less than most amateur astronomers," he said. "I can name twenty. The keenest astronomer could do maybe a hundred. He can name thousands."

"But that's incredible," I said.

"At one time memorizing things was considered a very important skill, as important as logical reasoning. Whereas now it's not valued at all because we just go to Google," Norris said. "The nearest thing I can compare Bill to is the ancient Greeks."

———

The Wardaman's traditional territory is bordered by rivers, the Daly Fitzmaurice and Flora Rivers to the north, the Katherine River to the east, and the Victoria River to the west. It is a tropical savanna of eucalyptus, bloodwood, ironwood, and whitebark trees, with rocky escarpments and gorges. From May until October, the region is dry and sunny, and then come November it is flooded with monsoon rain, and temperatures rise into the nineties. I arrived in Darwin to a wave of sticky humidity, but as I drove along the two-lane highway south into the interior of the Northern Territory, the air began to dry and the trees became less and less verdant. It was the month before the monsoon, and the landscape was so parched the trees were dropping their leaves to survive till the rain came.

Water is a central feature of Wardaman Dreaming stories, which tell of a Rainbow Serpent who brought the sea and flooded the land. Finally the willie wagtail bird, brown and peregrine falcons, and the Lightning Men—powerful spirits—conspired to kill him. They threw spears from Barnangga-ya, the top of Mt. Gregory, and chopped the Rainbow Serpent's head off. His eyeballs came out of his head and landed several miles away, creating two waterholes called Yimum. But even after he was killed, the land was easily flooded by rain, and so the black-headed python and water python got digging sticks and made all the rivers to hold the water, naming the places and creating songlines as they worked.

My nearly two-thousand-mile journey from Perth to the Northern Territory to see Harney on his cattle station—his family's ancestral land just west of the outback town called Katherine—was a bit of a shot in the dark. I had written him a letter that went unanswered and discussed my arrival in a short phone conversation with one of his family members in broken English, made more challenging by a weak phone connection. I hoped he would welcome a visit from me, but there was no way to be sure he even knew I was coming. In the meantime, I had read whatever I could about Harney's life and family, much of which came from his autobiography, *Born under the Paperbark Tree*. Many birth stories in Wardaman culture start with the surrounding plants or landscape features; the names of babies might include a gender prefix, the word for head, and the name of the nearest plant or tree, thereby connecting the baby's arrival into the world with the place it occurred. Harney was born under a paperbark tree, of the genus *Melaleuca*, whose layers of paperlike skin and green leaves were used for medicine, containers, and shelter. His father had fought in Europe in the First World War and gone to the outback to try and escape the trauma of battle. In 1932 the elder Harney was driving a donkey team to build roads in the Northern Territory and met Ludi Yibuluyma, a Wardaman woman. Ludi and her mother and father, Pluto and Minnie, worked with Harney to create a road from the pastoral cattle stations of Willeroo all the way to Victoria River Downs, a distance of 150 miles. The work took nearly four years. During that time, Harney and Ludi had two children, Dulcie followed by Bill Jr. Once Harney moved on for work, Ludi married a Wardaman man by the name of Joe Jomornji and the family lived on the Willeroo cattle station. Jomornji became the young Harney's stepfather, and when the wet season came and cattle work was impossible, all the workers would put away their clothes and return to the bush for months at a time. It was during these periods that Harney learned Wardaman hunting, law, and Dreaming stories from Jomornji and his grandfather.

The Australian government had an explicit policy toward "half-caste" children: removal from their Aboriginal families and placement in orphanages where they could be assimilated into white Australian language and culture. Harney's grandmother, Minnie, had a child taken away, as did her daughters; in each instance, they never saw their children again. In 1940, a police officer took Harney's sister, and she was sent to a home—he was spared because he was so small. From then on, Ludi did everything she could to keep him, hiding him from the police and even painting him with black plum juice mixed with charcoal to cover his light skin. This might be one reason why the elder Bill Harney never wrote about him: to do so would have given the welfare agency evidence to put him in an orphanage. In an oral autobiography published in 1996, Harney said that whenever his dad came to visit, he encouraged his mother to "show him your history and the story, keep the cultural side going."

Throughout his life, as a child in the bush to his work as a stockman and horseman on cattle stations beginning at the age of ten, Harney had to negotiate the borders of Aboriginal and white identity in the Northern Territory, a region considered by Australians to be the equivalent of the Wild West. It was a shockingly violent place ever since the first pastoral settlers had arrived in the 1850s looking for land. The Wardaman put up a resistance to the intrusion into their country, and in turn the colonizers used massacres, poisoning, and indentured servitude to subdue them. At one such massacre at a place called Double Rock Hole, Aboriginal women and children were shot and pushed off a cliff to their deaths. Ludi Harney's paternal grandmother was among them, and her dead body, along with her child's, was put in a heap with others and lit on fire. Somehow the child, Yiba-daba, survived and was adopted; he was Bill Harney's maternal grandfather, also known as Pluto.

I was told this story by Francesca Merlan, an American linguistic anthropologist, who has spent years studying in the Northern Territory and who first heard it in 1989 from Elsie Raymond, a relative of Harney's. "These guys lived in a part of the countryside where there was a quite violent contact history as the pastoralists took over the countryside," Merlan told me from her office at Princeton University, where she was a visiting scholar at the Institute for Advanced Study. "Their real purpose was to eliminate the Aboriginal people, if they could. Elsie and people at her age were just on the cusp of those massacres. Their parents in that generation were the immediate victims and survivors." Merlan went to the Northern Territory in 1976 as a young student with a grant to study the Wardaman language. Throughout the 1960s, Aboriginal people were thrown off pastoral "stations," and most lived in Katherine or in camps on the fringes of town. The stations' white owners tried to keep Aboriginal people away from their properties in order to prevent them from foraging, hunting, or conducting ceremonies, and they were hostile to Merlan's presence on the land as well. Her best sources were in town, where she found prolific storytellers, like Raymond, who could recount the history, genealogy, and mythology of the shrinking community. During these conversations and ventures into the bush, Merlan discovered most had an encyclopedic knowledge of the landscape. "We always moved around it with total confidence," Merlan told me. "They knew exactly where they were all the time, even if we'd walked for four days. No way would they have ever gotten lost." And those places where massacres had taken place were not forgotten; the memories were fresh. "Elsie and her father and mother had these endless stories of being shot at," said Merlan. "After awhile, you had to notice that this was a very deliberate strategy. They tried to kill everyone they could find."

As I drove from Darwin to Katherine I began to realize the landscape was unlike anything I had seen before. There were slight slopes and rock outcroppings, but mostly I drove through what seemed like a never-ending forest of sparsely spaced eucalyptus, or "gum" trees. For nearly two hundred miles, towering scarlet gum, poplar gum, ghost gum, river red gum, and white gum trees almost bare of their leaves passed outside my window. In between the trees, long yellow grass grew high except for intermittent patches where it was blackened by fire. To me, this singed landscape looked half dead, but Aboriginal people see burned landscapes with a sense of satisfied pride. Burning is beautiful because it is evidence that the land has been cared for and the creation of the ancestors maintained.

The historian Bill Gammage has written extensively on the practice of burning in precolonial Australia, describing fire as an ally for Aboriginal people, who worked the land "as intimately as humans can." Fire, he writes in his award-winning book *The Biggest Estate on Earth*, "let people select where plants grew. They knew which plants to burn, when, how often, and how hot. This demanded not one fire regime but many, differing in timing, intensity and duration. No natural regime could sustain such intricate balances. We may wonder how people in 1788 managed this, but clearly they did. The nature of Australia made fire a management tool." When European settlers arrived in Australia, the landscape they encountered was not wild in any sense. Its features—often remarked to be as beautiful as the gentrified English countryside—were the result of applied geographical and ecological knowledge, all encoded into the Aboriginal Law and Dreaming songs. Gammage argues that a key truth about the Dreaming is its fusion of theology and ecology: "In its notions of time and soul, its demand to leave the world as found, and its blanketing of land and sea with totem responsibilities, it is ecological. Aboriginal landscape awareness is rightly seen as drenched in religious sensibility, but equally the dreaming is saturated with environmental consciousness."

I arrived in Katherine at sunset and slept a few kilometers outside of town at the home of the local doctor, his wife, and two young daughters. Their open-air house sat on several dozen acres of bushland that stretched down to the Katherine River, and we ate dinner in the near dark. Around the corner from my bedroom was the open bathroom, and they warned me to shine my flashlight on the ground to avoid stepping on the olive pythons that slither onto the concrete floor of the shower to absorb its heat. I woke at dawn and stopped at the supermarket on my way out of town to buy provisions—bread, avocados, cheese, butter, and tea—and then set out on the Victoria Highway heading west. Harney's grandfather Pluto had helped to build this highway, leading a bulldozer along a Dreaming track while Harney, just a young boy, walked alongside listening to his grandfather's stories.

After about a hundred miles I saw an old oil drum can and a faded piece of wood with the word "Menngen" written on it. I turned right and stopped the car to open a cattle gate. The dirt road was soft and red, and the gum trees seemed even more barren. Over the next twenty miles all I saw were scattered cattle, broken-down cars, and a few mummified carcasses of dead donkeys lying in the grass. I opened and closed one cattle gate after another until I finally came around a bend to a grove of shady trees and houses surrounded by old machinery and horses in paddocks. No one seemed to be around until finally Harney himself pulled up in a baby blue pickup truck and greeted me. He had gotten my letter and seemed happy and completely unsurprised to see me. At eighty-three years old, he was smaller than I imagined, and his hair was grayer than in pictures I had seen. He wore a brown oilskin decorated with a thin braided rope and old cowboy boots, gray pants, and a plaid shirt. He told me he had gone into town that morning to get some beer, whiskey, and provisions

for our trip: he was going to show me one of Menngen's hundreds of rock-art sites, a Dreaming place, and then we would
camp near a waterhole for the night.

I grabbed my swag—Australian lingo for sleeping bag—and
my cooler or "Esky" and jumped in the passenger seat of his truck,
its armrests filled with old pens, combs, and canisters of chewing
tobacco. Harney got in the driver's seat and we began to drive
slowly through the maze of dirt tracks that penetrate the interior
of Menngen. As he drove, Harney began giving me a natural history lesson of the plants, animals, and landscape around us. In
compressed English, he enthusiastically described and narrated
all we observed, translating, in effect, the landscape for me. I was
reminded of what the archaeologist Isabel McBryde had written,
that any time she traveled with Aboriginal people she was "constantly impressed by how different were the landscapes . . . [we]
were observing. Theirs were numinous landscapes of the mind,
peopled by beings from an ever-present Dreaming whose actions
were marked by the features of the created landscape." I merely
saw trees, grass, and dirt bleached by heat and sun, while Harney
evidently saw a landscape teeming with history, food, medicine,
shelter, tools, and stories.

He named all the trees ("That's a coolabah tree, there's a wattle
tree, there's a stringing bark, that's bloodwood . . .") and how they
could be used by humans and animals. He pointed to the termite
mounds in between the trees, some four feet high or more, likening
their shapes to people or objects. "Looks like a little kid standing
next to them! Look at that one—that one's interesting!" Women, he
told me, use the termites for medicine after giving birth to ensure
that their milk will come in. "Put [them] in the fire, heat it up, crush
it up, make it like a powder," he said. "They get a special type of
grass, set the grass on top, steam comes through. Cook the breast."
We talked about his grandfather Pluto, who had helped to create
the Victoria Highway. "Grandfather walking, telling us the story of
all the different places, that how we grew up," he said. "Fly dreaming place, possum dreaming, pole cat, dove dreaming."

When Harney was barely a teenager, he began delivering mail by packhorse to cattle stations. "I was going up where there was story on the Aborigine walking track," he remembered. "From Willeroo all the way to Katherine, good walking track. Eighteen miles about four days. Just walking, no car, a lot of walking, remembering, singing. Took our time, weren't in a hurry to wake up any time in the morning." Can you sing those songs, I asked him, if you're driving in a car? "In a car when you travel, you miss a lot of country, a lot of stories," he said. Harney's knowledge of his country's stories was what enabled him to stake a claim to the land we were now traveling on. Merlan had told me how when she returned to Katherine again in the 1980s to continue her research and eventually publish the first and only grammar book on the Wardaman language in 1994, there were only about thirty people, all over the age of forty, who still regularly spoke the language. (Today, she estimates there are just a handful.) Around that time Harney was living in Katherine, working as a farmer and mechanic. He had two sons from his first wife, who had passed away from an undiagnosed brain tumor, and he had married a woman named Dixie. At that time Australian courts were beginning to recognize the ownership rights of Aboriginal people to their traditional country under the Land Rights Act of 1976. As Harney recounts in his autobiography, he told Dixie, "Well, everyone else is claiming the land. You follow me, I want a land claim on my country, back of the Flora [River]."

In order to claim ancestral territory, Aboriginal people needed to show that they were the descendants of those who lived on the land, had spiritual affiliations and responsibilities to it, as well as the traditional rights to forage. In order to fulfill these requirements to the satisfaction of the courts, anthropologists were often enlisted to gather the Dreaming stories and history of the country, as well as to create maps and surveys of the territories in question. The Australian anthropologists Betty Meehan and Athol Chase were working on a land claim for the entire Upper Daly region and approached Harney. Was he connected to the region on the back of the Flora? As he recounted in *Born under the Paper-bark Tree*, he said, "Yeah, that's my proper Dreaming there."

"You know all the history?" they asked him.

"Yeah, I know all the history."

"Oh well, it's up for the land claim. You want to land claim it?"

"All right," he said. "We've got all these children coming up, we must take them back to the bush to show them the history and their heritage in the country, their story, their singing, and do lots of ceremonies."

"Best to fight the government and get your land, otherwise they'll destroy all your Dreamtime story in the country there," Meehan and Chase said.

Harney showed the anthropologists his Dreaming sites, traveling by airplane, helicopter, and car. He sang traditional songs in front of the judge in court. While awaiting a final decision on the claim, the judge granted Harney and his family the right to live on the land even though the owner refused to give it up and even dropped poison bait to kill their dogs. During that period of waiting, Harney's mother, Ludi, passed away at the age of ninety-three. When the land claim was finally granted for some seventeen hundred square miles to be controlled by the Wardaman Aboriginal Corporation, Harney and his descendants were granted ownership rights to one hundred square miles. Harney called it Menngen, after his mother's Dreaming, the white cockatoo.

In Wardaman tradition, the white cockatoo is the bird that keeps an eye on the other important birds like diver ducks and wedge-tailed eagles, who are in turn tasked with ensuring that people are following traditional rules in anticipation of sacred ceremonies that begin in October at the start of the wet season. Together the birds are guardians of the law, making sure that secrets aren't divulged to the wrong people, that taboos aren't broken, that certain borders aren't crossed. In the night sky, the white cockatoo is represented by the star Fomalhaut, which appears in the northeast in late July, heralding a change in seasons. It is also a part of a celestial songline that begins with the Creation Dog in the north and stretches across the sky to include the stars of the Rock Cod, Eagle, the Big Law Place, Red Ant Doctor, White-Faced Grass Wallaby, and Catfish Law, and ends with the Bats in the southern sky. The Bats, the constellation the Greeks called the Pleiades, represent the children and teenagers who will be initiates. By following this celestial sequence of stars, one would have been able to navigate to the traditional place of the ceremonies.

Harney was initiated at the age of twelve, he told me. Joe Jomornji decorated him with his own blood and put feathers over his body, and he stayed up for three nights listening to the men dance with didgeridoos and clap sticks, and the women danced too (though for this he was covered with a blanket so he could only hear, not see, what was happening). After the third night, at dawn, he was held down and circumcised with a stone knife. After that he attended the ceremonies every year and learned more Dreaming songs. "I grew up with the song. We sit down and know exactly what name is our places where we're going. We sing and we happy," Harney recounted in the book *Dark Sparklers*. "We sing like we know where we stay at a house. We sing like we know exactly where all these places. We name them: the routes to travel, right to the end of the songs. Yeah, I sing one and I follow my Dreaming: and my Dreaming stops and another Dreaming goes on. No matter what happens we sing together."

After an hour of driving, Harney started "looking for the good shade" and stopped near a coolabah tree for lunch. He ate boiled beef on bread and we shared cheese and avocados before continuing to drive on the dirt paths for another hour. I had long ago lost track of our bearings. Harney talked about his first marriage, his education in the law, and the Dreaming stories belonging to his father and his mother. The forest got denser and the track ahead was invisible under the thick brush. We startled a group of animals—red kangaroos, wild horses, and the auspicious brolga, beautiful white cranes native to Australia—gathered around a small spring. "If I get close," Harney said of the brolga, "I'll sing a song for them." But, surprised by the truck, the animals scattered.

Just beyond the spring he stopped the truck and we began walking through a dense grove of trees, vines, and foliage. I didn't realize it at first but the vegetation was disguising a giant boulder, at least sixty feet high. As we circled the rock, one of its bare faces was revealed, across which was an ancient mural painted in red, white, yellow, and black ochre. The images stretched from the ground up to twenty feet or so. There was the underside of a frog with a gaping vagina whose contours mimicked the natural indentation of the rock, and a series of "lightning people" from whose heads extruded beams of light. Harney pointed above our heads to footprints imprinted in the rock. These he said were left by ancestors, who had walked this way in the Dreamtime. As I looked closer I recognized two different-size prints: alongside the adult-sized footprint was the much smaller stamp of a young child, and I felt a shiver. These footprints weren't meant to be interpreted as symbolic representations of ancestors: they are the literal traces of individuals who traveled here long ago. In this place, the Dreaming was as real as the Laetoli footprints preserved in volcanic ash in Tanzania.

We sat on two stones beneath the boulder. Around our feet were the detritus flakes of stone tools and spearheads. Harney began to tell a story in Wardaman, pausing every few minutes to translate it into English for me. He told of a time when people traveled all over the country, women on the left side and men on the right, to make a "huge big creation song, got to name all these places, painted all the young ones. All day everybody's happy," he said. "Oh it was a good time." Then Old Rainbow heard all the noise and people decided, "Let's kill that Old Rainbow." Hence began an hour-long saga of floods, battles, the creation of weather, earth, and sky. Harney told each intricate sequence and detail of the story, explaining the birth of the land itself. I was riveted, disoriented, swept along by a version of history whose cast was unfamiliar. I knew this was a simple story, probably reserved for young children and the uninitiated, but it hardly mattered to me.

That night we made camp next to a beautiful, pondlike water-hole surrounded by river gum trees. We drank beer and ate around the fire. I told Harney about my two-year-old son back in the United States, and he gave him a new name in Wardaman, Wajari, meaning "mother away, kid behind," which made him laugh uproariously. As it got dark and the stars emerged in the sky, he picked up two sticks to clap and keep a beat, and began singing—a saga of humanlike animal beings roaming the land. Sometime in the middle of the night after we'd fallen asleep I felt rain on my face and opened my eyes. But when I looked around all I saw were thousands of stars sparkling above. Nearby, Harney was also sitting up in his sleeping bag, quietly observing the sky. "Huh," he said to himself, as perplexed as I was by the rain that fell from a clear sky. The next morning we saw the marks in the dirt that the raindrops had left behind, proof we hadn't dreamed it all. After breakfast we sat under a wattle tree drinking smoky-flavored Billy tea and continued to talk—about ancestors, traveling, the Wardaman words for directions, survival techniques if you get lost in the bush, fishing, medicine, and the law. "Would you draw something for me?" I asked. I wanted to see the Dreaming tracks that cross Menngen. Harney took my pen and paper and sketched the border of the cattle station. Then he began to draw the contour of one Dreaming track after another. "This is Grasshopper Dreaming," he said. "This is Water Python Dreaming. This is Diver Duck People."

All together he named forty-two Dreaming tracks before running out of space on the page.

I was careful not to lose my notebook after that. I camped around Katherine for a few days and then drove back to Darwin and flew to Sydney to begin the nearly ten-thousand-mile journey back home. I considered the notebook a precious reminder of my glimpse into another world whose reality began to predictably fade as I went back to the familiarity of my life. When I looked at it, I was reminded of the paintings of Yukultji Napangati, a Pintupi woman from the Great Sandy Desert, who was one of the last Aboriginal people to "come in" from the bush in 1984 at the age of fourteen. Napangati's paintings are composed of thousands of lines oscillating across the surface of the canvas. The pictures seem to defy two-dimensional form—their surfaces move and undulate. One depicts the rock hole site at Yunala, west of Kiwirrkurra in the Gibson Desert, the same place where Jeffrey Tjangala, Yapa Yapa Tjangala, Fred Myers, and David Lewis had gone during their desert adventure. To Napangati, Yunala is where a group of female Dreaming ancestors once camped and dug for bush bananas and silky pear vines. When I stared at the painting and let my eyes slightly lose focus, the thousands of lines seemed to become a topographical map representing every inch of the desert. The painting is a painstakingly rendered visualization of a memory of a place.

What was the key to Harney's, Napangati's, and so many others' phenomenal memories? It is, I think, a matter of intimacy. We seek to know what we love. The journalist Arati Kumar-Rao writes of her time traveling with nomads of the Thar Desert in Rajasthan, India, whose memory of the land is passed on through words, place-names, songs, and symbols, that because they travel on foot, they map in inches, not kilometers. Aristotle thought that memories are imprinted on us like a seal in wax. According to him, sensory perceptions are initially received by the heart and then delivered to the brain, where they are stored. Later, the Latin verb for recollection was *recordari*, an amalgam of *revocare* ("to call back") and *cor* ("heart"). In the twelfth century, one of the meanings of the Middle English word *herte* was "memory." When you have memorized something completely, you know it by heart.

YOU SAY LEFT, I SAY NORTH

Does the language we speak and the words we use to describe space influence our perception of reality? The psychologist James Gibson describes knowing as an extension of perceiving. Children become aware by looking around, listening, feeling, smelling, and tasting, and others step in along the way to make them aware through instruments and tools, toys, pictures, and words. Yet all of these things—especially words—are different depending on what culture you are born into. Do people raised in different cultures see the same world?

Around the same time Gibson was studying psychology at Princeton, the American linguist Benjamin Lee Whorf was at Yale University trying to understand how the human mind might organize experience and phenomena differently depending on the linguistic system people used. Whorf studied Native American languages under his mentor Edward Sapir, and their body of work resulted in the Sapir-Whorf hypothesis—the idea that language determines and limits thought and knowledge. This theory was influenced by the work of Franz Boas, the German-American anthropologist who had lived on Baffin Island in the 1880s and who had noted the diversity of words the Inuit possessed to describe snow. (Sapir later became Boas's student.) While Sapir proposed that language influences how people think, Whorf went further, asserting that language reflects entirely different concepts about the world, including fundamental notions of time.

Sapir passed away in 1939 and Whorf died just two years later. Over the subsequent decades the majority of linguists, largely led by Noam Chomsky, adopted a different theory of language, that of universal grammar. The universalists believed that different languages each describe basic, innate concepts and that we are born with an a priori knowledge about things like space and time that are informed by our shared biology, the boundaries of our physical bodies, and our cognitive structure.

It was decades before the Sapir-Whorf hypothesis was revived, in part due to the work of an American linguistic anthropologist by the name of John Haviland, who lived with an Aboriginal community of about eight hundred people on the Cape York Peninsula in northeastern Australia. Haviland had gone there in the early 1970s to do postdoctoral research, and the first thing he did was begin to learn their language, called Guugu Yimithirr, from an old man called Billy "Muunduu" Jack. Eventually Jack gave Haviland a room in his home and adopted him as a son. As he learned the language, Haviland realized that while many European languages rely on phrases like "in front of" and "behind," these didn't exist in the Guugu Yimithirr speakers' grammar. Such phrases depend on an egocentric frame of reference—they use the position of the speaker or subject as the point of orientation. Sentences like "She is to the left of the tree" or "The tree is to the left of the rock" utilize a relative conception of space that depends on the perspective of the speaker to the tree or the rock. Guugu Yimithirr speakers did not have words for "right," "left," or "back." They exclusively used cardinal directions to describe space, which for them were four quadrants, skewed about seventeen degrees clockwise from magnetic compass directions. The use of these four root terms pervaded their speech and storytelling.

If instructing someone to turn off a camping stove, they would say, "Turn the knob west." If they wanted someone to move over, they would say, "Move a bit east."

In 1982, a British linguistic anthropologist at the University of Cambridge, Stephen Levinson, went to Australia to do a research fellowship at Australian National University. His interests were in psycholinguistics—the relationship between cognition and language—and in particular exploring the boundaries of language's power on human thinking. Haviland's findings fascinated him. Whorf thought that the human mind might organize phenomena uniquely depending on the linguistic system used, but his caveat to the idea was space. "Probably the apprehension of space is given in substantially the same form by experience irrespective of language," Whorf wrote. Now Levinson thought the Guugu Yimithirr community might prove otherwise, that those who use different languages to describe space might actually perceive the world differently.

In 1992, Levinson returned to Queensland to continue his studies and found that cardinal directions were so entrenched in the Guugu Yimithirr's perception of their surroundings that they even read books and watched television differently from Indo-European speakers. As he would write later, "[J]ust as we think of a picture as containing virtual space, so that we describe an elephant as behind a tree in a children's book (based on apparent occlusion), so GY speakers think about it as an oriented virtual space: if I am looking at the book facing north, then the elephant is north of the tree, and if I want you to skip ahead in the book, I'll ask you to go further east (because pages are flipped from east to west)." Levinson recalled an instance in which he showed ten men a six-minute film and then asked each of them to describe the events of the film to another person. He saw that if the men had been facing south while watching the film, they would describe people on the screen coming toward them as moving northward. "You always know which way the old people been watching the TV when they tell the story," one man told him.

Guugu Yimithirr speakers used what Levinson deemed an absolute frame of reference for spatial cognition: their perspective was from a fixed point of reference, rather than changing as they moved through space. In this scheme, their own bodies were sometimes treated as irrelevant. Guugu Yimithirr speakers would sometimes point at their chest, not to indicate they were referencing themselves but to point in the direction of something *behind* them. Levinson guessed that this kind of linguistic system could have far-reaching cognitive consequences. Perhaps Guugu Yimithirr speakers encode their memories with the cardinal directions. They would have to do this constantly, because people never know in advance what they will need to recall later. The result, he thought, was that Guugu Yimithirr speakers must have uniquely developed memory faculties from people who use egocentric frameworks for describing space—one difference being that Guugu Yimithirr speakers have to know the cardinal directions *all the time*. And this, he guessed, meant they could probably dead reckon with great precision. "Speakers of languages that utilize cardinal directions (like 'north'), where we would use coordinates based on our body schema (like 'left' or 'front'), would have to be not only good at knowing where (e.g.) north is," Levinson surmised, "but would also have to maintain accurate mental maps and constantly update their position and orientation on them."

To test this idea, Levinson turned to David Lewis's experiments in the Western Desert two decades before. He began traveling with Guugu Yimithirr men, ten in all, on journeys through the bush by foot or vehicle. He asked the men to indicate the direction of places like islands, cattle stations, and mountains from where they were standing, places that could range anywhere from a couple of miles to hundreds of miles away. As they were asked the men would point, often immediately, to the location requested of them and Levinson would then take a reading with his compass and record their position. He then compared the directions he was given to survey maps. All together Levinson gathered 160 readings, and the results were astounding. On average, the error in degrees from the men's dead reckoning and that of Levinson's compass was 13.5 degrees. "Nothing like this can be obtained from European populations," Levinson wrote in his book *Space and Language*.

Could it be that people who spoke languages that demanded constant dead reckoning were empirically better navigators than speakers of Indo-European languages? Based on his experience in Queensland, Levinson believed the answer was yes. But to prove it, there needed to be research beyond Australia to places where similar "absolute" spatial languages were spoken. Levinson organized a group of students whose goal was to research the domain of cognition, action, and space among different language groups around the globe, and in the following years, several studies emerged that reenacted these dead reckoning experiments.

In the Chiapas Mountains of Mexico, Levinson and his wife, Penelope Brown, an American psycholinguist, did field research among the Tzeltal Maya Indians, an indigenous group who live in a subalpine and agricultural setting of Tenejapa. The Tzeltal's language contains both absolute and relative terms. But the region the Tzeltal people inhabited was fairly small and they used well-maintained trails to travel through the mountains. When tested, their dead reckoning skills didn't match the Guugu Yimithirr's level. Those who relied on absolute language systems alone, it seemed, were indeed much better at orienting and therefore navigating.

One of the people who set out to re-create Levinson's experiments was a German anthropology student by the name of Thomas

Widlok. Widlok's interest was in the San—the hunter-gatherer group previously known as the "Bushmen" spread across Botswana, South Africa, Namibia, and Zambia—and in 1993 he went to northern Namibia to undertake an orientation study with the Hai‖om San (the "‖" represents a click sound used in San language, in this case made with the tongue placed on and then drawn away from the teeth.) The Hai‖om, a group of about fifteen thousand people living in the Kalahari Basin, and other San groups had a reputation for nearly mythical powers of navigation. Widlok had read the literature documenting these skills. One hunter claimed that his San guide's sense of direction was better than his handheld GPS device. Widlok also knew that during the border war in Angola and northern Namibia in the mid-twentieth century, the South African Army had created what was an elaborate ideology around these skills and exploited them to track down their enemies in the bush. The depiction of "Bushmen," Widlok has written, was that they were skilled superhumans, or rather super beastly beings. Indeed, white army officials, much like nineteenth-century anthropologists, projected animality onto the San. To outsiders, they were not people "partaking of nature, it is nature that has not yet completely released its grip on them."

Widlok knew from his experience that the Hai‖om could accomplish orientation tasks that seemed impossible to him, such as easily locating places they had never been before. But how much did the language of the Hai‖om play a role in these skills? Widlok took a GPS to the Mangetti-West region and began a study, accompanying six men, three women, and a twelve-year-old boy into the savanna, walking anywhere from nine to twenty-five miles. Widlok asked them to point to twenty different places ranging from a mile to over a hundred miles away. The visibility across the bush was around twenty yards, and there were no landmarks to be seen. Show me where X is from here, Widlok would ask, and then he noted the direction they pointed in and compared their estimates to his GPS reading. Again and again, Widlok found that the Hai‖om's dead reckoning skills were statistically barely different from the Guugu Yimithirr study group.

Widlok's data revealed something else. Unlike Levinson, he had included women in his study. In Hai‖om culture, it is the men who often attain expert levels of hunting and tracking skills, and yet he found that the women demonstrated even better dead reckoning than the male participants. Widlok guessed that this difference in gender might even out with a larger sample size. But there was an additional possibility. Western researchers long believed that gender was an important factor in spatial orientation and memory, and studies had shown that on the whole men performed better than women in wayfinding and spatial ability tasks. Indeed, as Indiana University psychologist Carol Lawton has pointed out, spatial abilities studies are often used by psychologists to show evidence of gender differences in cognition, mainly because differences are otherwise so minimal. They point to results that have shown that boys score higher than girls on mental rotation tests—judging how an object would look if turned—and other exercises like orienting in space. It's only when psychologists test girls on their memory for locations of objects that they exceed boys. Various hypotheses for these gender differences have been presented, from hormonal differences and their effects on our hippocampus, to how our brains are hemispherically organized, to evolutionary causes. Perhaps men in our ancient past needed to travel farther to hunt, search for mates, or fight, whereas women were circumscribed by gathering food and protecting their offspring. But as Lawton points out, there's no clear evidence that labor was divided in this way in prehistoric times. And interestingly, gender differences in navigation seem to disappear when boys and girls from lower socioeconomic groups are tested. Furthermore, when women are offered instruction and practice in spatial visualization exercises, the differences between their abilities and men's disappears. One study by Ariane Burke of the University of Montreal tested the hunter-gatherer theory of spatial ability and found that men and women with equivalent experience

perform equally well at navigation tasks once physical differences are accounted for.

The research into absolute languages seems to indicate that
gender difference might be culturally dependent rather than gender dependent. Perhaps women who spoke languages with absolute spatial coordinates, thereby constantly orientating in order
to communicate, were as capable as men. This hypothesis gained
some traction when, in addition to the Guugu Yimithirr, Tzeltal,
and Hai‖om experiments, Dutch and Japanese studies were undertaken that tested dead reckoning skills. Only the Dutch study
group, whose language relies almost exclusively on relative, egocentric terms and whose participants struggled to access an absolute frame of reference, showed a difference between men and
women.

Like the Guugu Yimithirr, the Hai‖om had no general words
for "left" or "right," though those who could speak English were
perfectly capable of employing the words correctly. As it turns
out, so many cultures use this absolute frame of reference rather
than an egocentric perspective. Almost all Australian Aboriginal
language groups, not just the Guugu Yimithirr, use an absolute
linguistic frame of reference. So do Dravidian speakers in India;
Totonacan speakers in Mexico; and Balinese speakers in Indonesia. And it's not always one or the other. Some groups use both
frames of reference, like Kgalagadi speakers in Botswana and Kili-
vila speakers in Papua New Guinea.

The reasons for the proliferation of linguistic diversity when
it comes to space and orientation are not fully understood. Do the
environments we inhabit shape our languages and therefore cognitive architecture as we grow and develop? There's no proven
deterministic relationship between what ecologies and cultures
lead to relative or absolute frames of reference, but there are relationships. In a 2004 study in the journal *Trends in Cognitive Sciences*, researchers looked at ten different communities and found
an association between living in urban environments and using
a relative frame of reference, and living in rural environments and

using an absolute frame of reference. In some cases, however, a rural community relied on a relative frame of reference, such as the Yukatek, a rural community in Mexico. What does seem to be generally true is that hunter-gatherer societies almost all seem to use absolute systems.

What was clear to the authors of the 2004 study was that language used by different cultures to describe space proves that there are no innate spatial concepts built into human cognition. The relative frame of reference described by Immanuel Kant was not somehow more "natural" to humans. In fact, children acquiring relative languages seem to do so with greater difficulty and at a much later time than absolute language speakers. English, Italian, and Turkish children can't use relative descriptors like "right" and "left" confidently until eleven or twelve years of age. On the other hand, Tzeltal-speaking children can use absolute vocabulary by three and a half and have fully mastered it by eight. "Language," wrote the researchers, "can play a central role in the restructuring of human cognition."

———

When it came to explaining the Hai‖om skill for tracking and navigating, Widlok knew there were two very different, contradicting theories within his discipline of anthropology: mental map theory and practical mastery theory. The first posits that successful navigation relies on building abstract cognitive representations of the spatial relations between objects in the mind, while the second asserts that successful navigation is a matter of memorizing the perspective of moving along a route. One relies on survey or layout knowledge, while the other relies on sequential knowledge. In his 1997 paper discussing the results of his field research, Widlok seemed to place the Hai‖om system of navigation in the practical mastery camp. Theirs wasn't a system based on a map against which their movements were referenced and direction oriented. Yet Widlok also felt that the Hai‖om had an extraordinary amount of diversity when it came to choosing which tools to use to both describe and orient themselves in space. These tools weren't an automatic response to environmental stimuli, Widlok wrote; they were constituted through "prolonged social interaction." For instance, they used what is called a "!hus" system, a way of classifying various ecological landscapes and the different people who lived in each one. There were stony-ground people, hill people, millet people, soft-sand people, and fine-sand people. These landscape terms were a way of talking, thinking, and traveling through space that took into account multiple dimensions of social history, individual memory, and ecological knowledge.

Widlok also noted how language itself and the Hai‖om speakers' use of spatial descriptors in conversation reinforced people's orientation skills. The Hai‖om were constantly describing space to one another; Widlok calls it "topographical gossip." This made sense: there were no material maps, and so the Hai‖om communicated in an almost continual flow throughout the day to share information about the locations of places, people, stories, and resources. When they talked about their travels to these places, they used these !hus landscape terms as cardinals. And yet none of the terms used focused on the egocentric body. The Hai‖om oriented geocentrically using a shared spatial language. Did this mean they were utilizing mental maps? Widlok was suspicious of the mental map theory. Academics who attempted to understand non-European methods of orientation seemed unable to transcend the concept. They were, Widlok told me, "overmapped." Maps require their users to use what is called nonindexical information (what the map tells them to expect) with indexical information (what they actually see), meaning navigation with them is a constant process of comparing where you are against a map, whether it is physically in your hands or in your mind's "eye." This wasn't like what he observed among the Hai‖om. Oftentimes landscape categories were combined with an individual's knowledge of botany, topography, intergroup relations, and individual life history and memory; all came into play as they traveled. So while they could dead reckon their distance and speed with great accuracy, they also used a whole lot of information, including a "patchwork of landscape countries that intersect with commonly used routes between named places."

I talked to Widlok on the phone at his office at the University of Cologne, where he is a professor of African Studies. He explained how beginning with Franz Boas in the Arctic, a significant revolution had taken place in anthropology; instead of explaining differences between people as the result of biology, anthropologists emphasized culture. But what exactly *is* that?

Widlok's work is part of a second revolution that has revisited the very idea of culture and how it commingles with individuals. "We've moved away from the container model of culture, the idea that as an individual we are part of a singular culture that over-determines what we are doing," he said, "and moved on to practice and social relationships. Whatever skills we have or the San have, it doesn't have to do with racial difference, nor is it so simple as their culture or language that determines what they are doing. We have lots of evidence that people move in and out of these systems of knowledge and we can combine them." This social theory of culture, or what some call a relational or practice-oriented theory of culture, emphasizes that human knowledge and culture are founded in skill, ways of doing things, interactions, habits of life, learning, and embodied practice—they are generated from the individual entangled in relationships. Culture is a process; not something handed down to us as much as something we re-create through our own participation.

The idea that culture is at the heart of cognition—and that cognition is itself culturally relative—implies that Western philosophers and scientists have long mistaken their own cognitive traits as universal. "The spectrum was widened," said Widlok. "What many people didn't consider to be possible was all of a sudden part of the spectrum of what humans and societies have come up with." Once we find evidence of diversity in human spatial cognition, it calls into question other assumptions that have been made about the human experience.

———

Anthropologists still disagree on the theoretical explanation of how humans engage in day-to-day wayfinding, a split between the mental map theory of navigation and practical mastery theory. As Kirill Istomin, an anthropologist at the Max Planck Institute, and Mark Dwyer, a geographer at the University of Cambridge, have written, "The central point of debate between these two theories is whether mental maps exist and, if they do, whether they explain human spatial orientation."

Anthropologists who subscribe to the practical mastery theory believe the wayfinder relies on visual memory and is immersed in cultural practices, habits, knowledge, and the direct perception of the environment. The core of this idea comes from a French sociologist by the name of Pierre Bourdieu. In the 1970s, Bourdieu published a book in which he argued that mastering a spatial environment arose through familiarity with "practical" rather than "Cartesian" space. Practical space, he said, was built through the perceptions and activities of the individual. Cartesian space, on the other hand, was the absolute spatial relations between objects, independent of the viewer. In 1985, British social anthropologist Alfred Gell, the same person who wrote about traps and human evolution, used these ideas to form the practical mastery theory of navigation, which refuted the existence of maps in the mind.

To understand the debate, I spoke to a proponent of the practical mastery view, Tim Ingold, the chair in social anthropology at the University of Aberdeen. Ingold believes that wayfinding isn't about abstract representations of spatial relations but occurs from the perspective of a path of observation. His definition of wayfinding is "a skilled performance in which the traveler, whose powers of perception and action have been fine-tuned through previous experience, 'feels his way' towards his goal, continually adjusting his movements in response to an ongoing perceptual monitoring of his surroundings." The Australian Aboriginal traveler wayfinds in this way across the desert just as the Micronesian sailor does across the open ocean and the Inuit dogsledder does when crossing the sea ice.

Ingold explained that his interest in the concept of wayfinding had roots in his childhood, when he was captivated by the polar north, books about Arctic explorers, and stories of grand adventure to unknown places. When he set out to do anthropological fieldwork, he lived with the Sami in Finland as a doctoral research student, taking notes on kinship, household economics, and the Sami's adaptations to the environment. They were a people, he quickly realized, that moved constantly. While they had permanent dwellings, their life seemed to exist completely outdoors along the extensive routes that they followed to locate their reindeer. He soon found out that they navigated by memorizing sequences of natural landmarks like trees, hills, swamps, and rocks, and the names associated with significant places. They would often orient by the direction of streams and rivers, and chains of hills. When the sun was unavailable or the stars were invisible, they would use tree branches and anthills to find north and south. Memorizing the *sequences* of their movements, Ingold realized, was important because the Sami could then reverse the sequence to find their way back and to relate the journey to others, to tell the *story* of their journey. In a paper Ingold coauthored with Nuccio Mazzullo, they described the Sami way of being as "not in but along. The path, not the place, is the primary condition of being, or rather of becoming."

According to Ingold, it was those early experiences of traveling with the Sami that first got him thinking about wayfinding, although some of the significance of what he observed and learned took many years to manifest. "Wayfinding's not something that people explicitly talk about. When I was doing fieldwork with the Sami, I wasn't thinking about these things at all," he told me. "I think it took decades to actually realize what I'd learned, why

I was thinking in this way and not that way." Later, he began to see how people move and dwell in space as an essential aspect of being, and that the diversity of these behaviors challenged essential Western assumptions about the divisions between nature, society, and people. When he discovered Gibson's *The Ecological Approach to Visual Perception,* it was a revelation. Gibson's ideas meant that perception is not the result of a mind in a body but of an organism as a whole in its environment, and is "tantamount to the organism's own exploratory movement through the world." Ingold saw this was a way of bringing together the biological life of organisms and the cultural life of the mind in society. We are not self-contained individuals confronting a world *out there*, but developing organisms in an environment, enmeshed in tangled relationships. As we move through space, our knowledge undergoes continuous formulation; wayfinding isn't knowing *before* we go, but, as he put it, "knowing as we go."

Lately, Ingold told me, he's reconsidered the word *wayfinding* and prefers the word *wayfaring*. It's an attempt to move away from the concept of navigation or mere transport to describe human travel. "I just wanted to get away from the idea of going from A to B. Wayfaring is the compound of two words. *Way* is like talking about a way of life, carrying along a life," he said. "In English, the word *fare* is a lovely word. How are you faring? How are you getting on?"

For Ingold the notion that aspects of the structure of the world are copied onto an analogous structure in the mind in the form of a map, which is then constantly updated, is nonsensical. It not only fails to capture the dynamic complexity and the skills involved in finding one's way through a landscape as he witnessed in places like Sami country ("an immensely variegated terrain of comings and goings, which is continually taking shape around the traveler even as the latter's movements contribute to its formation"), but it also falls short because maps themselves are a cultural invention. Indeed, the great myth of cartography is that maps are independent of any point of view and equally valid wherever one stands when, in fact, they always take a point of view, prioritize information, choose a scale. In contrast, Ingold argues that extending the map metaphor into the domain of cognition ignores the wisdom and practical sensibility of the navigator, divorcing tradition from locality, culture from place, and traditional knowledge from the environmentally situated experience of practitioners. It is, in other words, not how most humans actually experience the world.

Ingold draws a distinction between mapping, which is what wayfinders do, and mapmaking, which is what cartographers making physical maps do. Mapping is an act of committing to memory the experience of bodily movement and reenacting it. It's a kind of performance, like telling a story. On the other hand, in his book *The Perception of the Environment,* he writes that the mapmaker has no need to travel at all.

> [I]ndeed he may have no experience whatever of the territory he so painstakingly seeks to represent. His task is rather to assemble, off-site, the information provided to him—already shorn of the particular circumstances of its collection—into a comprehensive spatial representation. It is of course no accident that precisely the same task is assigned, by cognitive map theorists, to the mind in operating upon the data of sense.

Ingold has described wayfinding as an act of remembering similar to how one remembers a piece of music. Just as with musical performance, wayfinding has an essentially temporal character: "the path, like the musical melody, unfolds over time rather than across space." It struck me as not so different from Howard Eichenbaum's description of navigation as hippocampal networks that encode "journeys through space as memory episodes defined by sequences of events and the locations where they occur."

Why, I asked Ingold, was the debate over practical mastery and mental maps so divisive? "People on each side can hardly make any contact with one another at all," he lamented. "In a heated argument you have to have enough common ground to stage a debate. But I find with the cognitivists that they are in their own world. It comes down to what we mean by space in the first place. I think space stands for the possibility that there can be very many stories. It's like music. Space is the simultaneity of all these stories going on at once."

I asked Widlock the same question. "It's important to have Ingold's perspective as a corrective to the Western perspective," he said, which often treats all spatial knowledge as maplike. "But the main point to me is that humans are able to switch *between* perspectives. Even the San and others are doing that. They too can stand back and theorize and engage in different modes of thinking. Evolutionarily, the particular skills of humans are to *manage* very different perspectives. The San go tracking, and they do have the flow of perception and picking up things in the environment. But then they also stand there and compare hypotheses and practice rational, detached thinking."

Widlok's point seemed important for understanding human navigation, but also intelligence itself. Perhaps it is the flexibility of our thinking—of assuming different viewpoints, embodied experience, and abstraction—that has made us so unique. And perhaps our capacity for empiricism, inference, and theorizing is far older than we think.

PART THREE

OCEANIA

EMPIRICISM AT HARVARD

In the Western version of history, the origins of science are found in the work of ancient Greece's natural philosophers, whose torch of rational inquiry was picked up by Copernicus, Galileo, and Descartes and ignited the Scientific Revolution. The Greeks and their progeny are thought to have possessed an intrinsic capacity for scientific thinking, a thirst for the accumulation of knowledge, and the desire to understand the world in its entirety. According to this hegemonic view, it was the intellect and courage of these geniuses that allowed them to transcend superstition and myth. Some historians of science today concede that other cultures have made scientific contributions, but the classic narrative remains that science is distinctly Western. Other cultures can become modern and scientific, but only the West *invented* those things.

What if science wasn't invented twenty-five hundred years ago, but hundreds of thousands of years ago, and the practice of navigation had something to do with it? In 1990 Louis Liebenberg, a white South African who frequently traveled with the San of the Kalahari, published a book titled *The Art of Tracking* in which he called tracking the "oldest science." According to Liebenberg, the twenty-first-century physicist seeking to formulate an accurate

hypothesis about the behavior of particles based on a combination of his own theoretical models and hard data isn't far removed
from the hunter-gatherer reading tracks in the ground or observing the weather to successfully hunt, and there is scant difference
between the intellectual abilities of the two. "[Tracking] may
well have been the first creative science practised by the earliest
members of anatomically modern (a.m.) Homo sapiens who had
modern intellects," he writes. "Natural selection for an ability to
interpret tracks and signs may have played a significant role in
the evolution of the scientific intellect." Liebenberg, a member of
Harvard University's Department of Human Evolutionary Biology,
thinks that trackers

> have to create a working hypothesis in which spoor evi-
> dence is supplemented with hypothetical assumptions
> based not only on their knowledge of animal behaviour,
> but also on their creative ability to solve new problems and
> discover new information. The working hypothesis may be
> a reconstruction of what the animal was doing, how fast it
> was moving, when it was there, where it was going to and
> where it might be at that time. Such a working hypothesis
> may then enable the trackers to predict the animal's move-
> ments. As new information is gathered, they may have to
> revise their working hypothesis, creating a better recon-
> struction of the animal's activities. Anticipating and pre-
> dicting an animal's movements, therefore, involves a
> continuous process of problem-solving, creating new hy-
> potheses and discovering new information.

The undeniable difference between the tracker and the modern physicist is scale. Liebenberg points out that the tracker's
knowledge is limited to each individual's observations, which are
transmitted and shared through oral tradition, while the modern
scientist has nearly instantaneous access to a huge body of knowledge through libraries, institutions, databases, and instruments

such as computers and satellites that extend their access and reach to information. But the distinction is, he thinks, technological and sociological rather than intellectual. "The modern scientist may know much more than the tracker. But he/she does not necessarily understand nature any better than the intelligent hunter gatherer," he writes.

I found this to be a fairly radical notion, but in Liebenberg's view the notion that rational scientific thinking didn't originate with the ancient Greeks but with hunter-gatherers actually solves a galling mystery in human evolution studies—why is it that despite great technological and societal leaps over the last ten thousand years, the hominin brain stopped growing in size long ago? In the previous several million years, our brains evolved in both size and neurological complexity, reaching an apex a few hundred thousand years ago. And then the growth stopped. One of the ways to reconcile this historical paradox is, as Liebenberg puts it, to assume that "at least some of the first fully modern hunter-gatherers were capable of a scientific approach, and that the intellectual requirements of modern science were, at least among the most intelligent members of hunter-gatherer bands, a necessity for the survival of modern hunter-gatherer societies."

I went to Harvard University to talk to another academic whose own ideas have converged with Liebenberg's in interesting ways. John Huth is an experimental particle physicist whose experiences practicing and teaching traditional navigation to students have led him to believe that navigation and its cognitive demands had a lot to do with the human invention of science. I took a Monday morning Amtrak train and arrived on campus just before the start of Huth's class "Science of the Physical Universe 26" (SPU:26). As I walked across the yard to the university's science center, there were still boulder-sized chunks of ice left over from Boston's record-breaking winter storms, but the tips of the trees swayed in the wind, laden with new buds. I passed through the revolving door of the science center and past clusters of undergrads into Hall D, an aging amphitheater with green upholstered seats and faded violet carpets. Taking a seat in the back, I spotted Huth in front of a lectern, surrounded by an array of gadgetry and scientific detritus. Wearing well-worn hiking shoes and a button-up shirt with the sleeves rolled to his elbows, he looked like he bears his laurels lightly.

Huth was an early project leader at the European Organization for Nuclear Research (CERN) Laboratory in Geneva, Switzerland, when it began building the Large Hadron Collider (LHC), the twenty-seven-kilometer-long underground particle accelerator. His focus was the ATLAS Experiment, a proton-collision detector that is described as one of the largest, most complex scientific instruments ever constructed. Along with three thousand other scientists, he helped to construct and operate ATLAS's detector and analyze the bewildering amount of data (enough to fill one hundred thousand CDs per second) created by the collision of high-energy protons traveling at near light speed inside the collider. The LHC is sometimes called the Big Bang Machine because one of its goals from the start was to locate the mysterious Higgs boson, a theoretical particle that makes up an omnipresent but invisible field that gives atoms their mass and could help explain the beginning of time. In 2012, an experiment Huth collaborated on at ATLAS helped find the Higgs boson, an occasion that marked what many described as a new frontier in the human understanding of matter. ATLAS's future work will likely focus on trying to solve mysteries like the origin of mass, extra dimensions of space, and black holes, and will search for evidence of dark matter. "The discovery of the Higgs was this crowning moment, and it's also the start of physics moving into unknown territory," Huth told me later. "We're in terra incognita now."

I was interested in how Huth came to teach an undergraduate class on traditional navigation without the aid of instruments and eventually to publish a book on the subject called *The Lost Art of Finding Our Way*. In August 2003, Huth told me, he rented a sea kayak in the Cranberry Islands in Maine. He steered offshore but a dense fog moved in and cut off his vision in all directions except for a patch of blue sky above. Huth hadn't prepared for anything other than a quick paddle; he had no compass, phone, map, or supplies with him. To keep from panicking, he forced himself to pay attention to his surroundings, noting the direction of the wind and the swell. He could hear waves breaking on a beach somewhere and he figured it had to be to the northwest based on his vague internal image of the coastline. He tried to keep track of time and judged his progress against what he imagined his position was in his mind's eye. Lobster buoys created little wakes from what he knew to be the incoming tide, and he used them as guideposts, finally managing to reach the shore.

This experience alarmed Huth enough that two months later when he was kayaking near Nantucket Sound off the coast of Cape Cod, he made sure he took note of the surrounding area before setting off. That day a fog also set in, and when it did he oriented himself by the direction of the wind and waves. Unbeknownst to him, less than half a mile away, two young women were also sea kayaking and enveloped by the same fog. When they didn't return as expected, the Coast Guard launched a search effort. The following day one of their bodies was found; the other was never recovered. Huth was devastated by this event. He guessed that the primary reason he survived and they didn't was his ability to interpret some of the environmental cues around him in order to orient in the right direction. It was possible that the women had misinterpreted their direction and paddled out to the open sea.

After that, Huth began amassing practical information on how to survive being lost without instruments, whether in the forest or at sea. He studied research on the behavior of people who are lost and on navigational practices around the world. One thing repeatedly struck him. Whether it was children in Greenland learning rope tricks for sea kayaking or Kalahari hunters tracking animals, navigation was evidence of an incredibly sophisticated body of science in cultures that were widely dismissed as primitive and without scientific traditions. "These cultures created a taxonomy in which they practice these arts," he told me.

"Navigation was a way of thinking or organizing one's environment in a scientific way. It's an example of scientific thought that people were engaging in way *before* the Scientific Revolution." The original impetus for his research had been a practical one: even the most basic knowledge can help someone survive unexpected circumstances outdoors, and he wanted to offer a synthesis of this information to students. SPU:26 was designed to teach them how to read the sun, stars, shadows, waves, tides, and currents in order to find their way, and to cover aspects of the navigational cultures of Polynesian, Norse, Arab, and early Western seafarers.

Now I watched fifty or so students file into the amphitheater and Huth begin his lecture. "Up to now what we've done is talk about navigation in various guises," he said. "Dead reckoning, stars, sun, using a compass. Now we're going to talk about the weather. Did anyone happen to notice the direction the wind was blowing before they got inside?"

No one raised their hand.

"Which way is the wind blowing? Anyone notice?"

"Southeast?" said a male student tentatively.

"Why do you think that?" asked Huth.

"Because the wind seemed to be at my back?"

"The wind is coming from the northwest," said Huth. "Around this area, buildings can make wind currents swirl, so the best way to note the wind direction is by looking up at the clouds." Huth began to scrawl across the chalkboard, explaining the mechanics of cloud formation, air density, wind, and geography. If you think this seems kind of elementary for a Harvard lecture hall and a particle physicist, you're not alone—I thought the same thing. Yet I couldn't have answered Huth's question about the wind. I had noticed a breeze on campus, but I failed to observe the direction it was coming from, not just that day but ever, because I wasn't in the habit of doing so. My method for observing the weather was the same as that of most people: I might glance out the window to decide whether or not I need a sweater, but to get a forecast, I turn to a phone or computer. I trust the forecasts I find there to be accurate, but for all I know about interpreting weather data, they might as well be handed down to me by Zeus and Hera. It's an example of what anthropologist Charles Frake calls "magical thinking." The example Frake uses is of tides, and the differences between how medieval sailors and modern sailors understood them. Modern Western society knows vastly more about tides, but tidal theory—based on complex mathematics—is beyond the understanding of individual navigators today. "Sailors today have no need to understand tidal theory at any level. They merely consult their tide table anew for each voyage. They never have to think about the tides as a system," writes Frake. "Consequently it is the modern literate sailor, not the medieval one, who is prone to 'magical thinking' about the tides."

Considering the students' average age was about ten years younger than me, I hazarded to guess that everyone else was equally technology dependent, if not more so. Clearly there had been a hemorrhage of basic knowledge of weather prediction to the point that skills that would have been familiar to many just a few generations ago were not widely practiced now. "In our era, people rarely notice the signs of the weather," Huth explained to me later. "But at a not-too-distant time in the past, weather dictated travel. Travelers had to cast their lot with fate or rely on their ability to read the signs in the clouds and winds to predict the weather for themselves."

Huth sees the willingness among his students to place their faith in technology and relinquish the power of their own observation as a larger, problematic trend in education. Students learn biology, chemistry, and geology—the result of hundreds of years of scientific discovery—but they atomize this knowledge rather than find a home for it within a larger conceptual framework, namely their own direct experience. And so, over the course of a young person's life in the classroom, students are surrendering the deeper meaning of what's taught to them to what Huth calls "guardians of knowledge." Their ability to see personal meaning and an existential stake in the environment around them disappears.

To illustrate how discouraging and pervasive this trend really is, Huth often cites a 1987 documentary called *A Private Universe*, made by the Harvard-Smithsonian Center for Astrophysics. The film's producers went to a Harvard commencement ceremony and asked faculty, alumni, and seniors, "Why is it warm in the summer and cold in the winter?" Of the twenty-three people questioned, only two could answer the question correctly. The film's message is clear: even the highly educated lack rudimentary knowledge about the environment they inhabit. Why do the seasons change? Why does the moon go through phases? Two hundred years ago, Huth points out, any bumpkin farmer—even if they hadn't gone to school and learned the earth rotates on an axis—would know that it is warmer in the summer because the earth is getting more direct light. "All empiricism has to start with stuff that is immediately palpable to you," Huth told me. "The march of education—especially in the sciences—has been divorced from reality and I think that's where you have to *start*."

Asking a classroom of elite students to pay attention to which way the wind is blowing, I realized, is a deceptively simple request. In order to succeed in Huth's primitive navigation class, students have to reclaim the most fundamental aspect of their experience, one that is so often mediated and cushioned: inhabiting space and time. It wasn't always this way, a truth brought home by the fact that at the other end of the building from where Huth teaches you can visit the university's Collection of Historical Scientific Instruments. In it are thousands of Western instruments for observing and navigating, like telescopes, sextants, compasses, and celestial globes, some of them dating back to 1400. These tools were integral to Harvard's core curriculum of "natural philosophy" when the school opened in 1636.

Huth's thinking about the significance of navigation has evolved since he had his kayaking experiences. The students he meets often come to class with a yearning to fill the gaps in their own knowledge. It was like the course slaked a thirst. By the end, a handful of students would relate heady, pseudomystical insights. "I'm leaving this class empowered to learn and simply be present," wrote one. "This course isn't about navigation but about a mindful way of living and finding our way and ourselves," said another. "To truly understand our surroundings we must immerse ourselves in them," wrote a third.

One of Huth's teaching fellows, a twenty-three-year-old PhD candidate in physics named Louis Baum, told me that he and the other fellows talk this way sometimes. "We get philosophical about it, how knowing where you are helps you know your place in the world," said Baum. "I find it comforting to be orienting myself. I notice it a lot among my friends: they memorize paths and then they don't pay attention." What Huth now thinks is that learning navigation ends up making students feel less isolated. "It's a framework that they use to put down observations, and it forces them to be keenly sensitive to their surroundings," he said. "And sometimes they're engaging in it and then experiencing an epiphany to other aspects of their life." The more attuned they are, the more their consciousness seems to expand. In this respect the effect of learning navigation struck me as echoing the discovery of a religious worldview or a transformative life experience: it thins the barriers between ourselves and the world.

———

In the summer of 2015, a few months after I visited his class, Huth joined a three-week expedition to the Marshall Islands, the chain of small volcanic islands and coral atolls in the South Pacific Ocean. Also on the trip was Gerbrant van Vledder, an expert in wind and wave dynamics and professor at the University of Delft, and Joe Genz, an anthropologist at the University of Hawaii who had spent years living and studying on the Marshall Islands. They had been invited there by the islands' traditional navigators,

who asked the scientists to join a voyage between atolls and collect
data and information that would shed light on their practice of
orienting and piloting over long stretches of open water using only
waves. For these navigators, the patterns of reflected and refracted
waves off land act as a kind of map. The patterns were traditionally taught to initiate navigators with "stick charts," weavings made
from plant fibers and shells that show lines and curves representing
the behavior of waves around islands. In the late 1800s, European
explorers had collected several of these charts and sent them to
museums back home, where they have been a source of cartographic fascination ever since.

Huth and the rest of the expedition hoped to answer long-
standing questions about Marshallese wave piloting. Genz has
described wave piloting as a "spatial field of multitudes of waves
coming from every possible direction, that [navigators] filter out,
it's an embodied kind of experience." But previous expeditions
to explain *how* had fallen short. Marshallese navigators could
describe reflected waves that were too weak to even be detected by
wave buoys, while other patterns they described seemed to con-
tradict the accepted wave transformation progress described by
oceanographers. How Marshallese navigators understood the
behavior of these extremely complex wave patterns, let alone "read"
or sensed them with enough accuracy to sail over miles of open,
seemingly undifferentiated ocean and make landfall at extremely
small atolls and islands, was a puzzle. The team of scientists, including Huth, armed themselves with instruments, computers,
GPS, compasses, anemometers, and satellite data and flew to the
Marshalls to find a connection between two seemingly dissimiliar
ways of seeing the world, indigenous and scientific, sensory and
technological.

For hundreds of years the outside world has been fascinated
by how people inhabited the South Pacific, a twenty-five-million-
square-mile area of ocean dotted with thousands of small islands.
How did humans possibly find their way to these small pinpricks of
land scattered across a vast and bewildering ocean three times

the size of Europe? As early as 1522 the historian Maximilian Transylvanus wrote that the Pacific was "so vast that the human mind can scarcely grasp it." When the Dutch explorer Jacob Roggeveen stumbled across Rapa Nui, a tiny volcanic island on the southern end of Polynesia, on Easter Day in 1722, he decided that the only way people with such small canoes could be living there was if God had created them separately from the rest of humanity. Julien Crozet, a French explorer, hypothesized that there must have been an entire continent of similar language speakers that had sunk below the water, leaving only the atolls and islands of the South Pacific above water.

For islanders in the South Pacific, Western explanations for their presence in Oceania were misguided and often insulting. In the 1940s, a New Zealand historian by the name of Charles Andrew Sharp claimed that humans couldn't sail more than three hundred miles out of sight of land without instruments—the maximum range, he believed, before navigational errors would doom them. As a result, Sharp thought distant islands must have been settled by unintentional voyagers; sailors probably fleeing famine or conflict being blown by storms or simply straying off course. His book *Ancient Voyages of the Pacific* put forth the idea that Oceanic islanders must have stumbled onto their islands from Southeast Asia by chance, accident, or fortuitous providence rather than conscious decision-making and skill. Just a few years before Sharp published his book, the Norwegian explorer Thor Heyerdahl famously sought to prove a similar theory, though he believed that these accidental colonizers of the Pacific had arrived from South America. In 1947 he launched a raft with six others from Peru and spent 121 days riding it along the prevailing east to west winds before making landfall in the South Pacific. Recounted in his internationally bestselling book *Kon Tiki*, Heyerdahl's ideas about the accidental peopling of Oceania became globally known—despite the fact that no South Pacific Islander agreed with him. As the anthropologist Ben Finney has written, "[S]tandard histories of cartography focused on physical map artifacts have largely ignored the way Oceanic navigators mentally charted the islands, stars, and swells." Nāʻālehu Anthony, a Hawaiian navigator, put it more bluntly: "All these things they told us about drifting off course was a lie. This was intentional. We did it thousands and thousands of years before anyone would lose sight of land in Europe."

Indeed, the Marshall Islands were first inhabited some two thousand years ago by people from the Eastern Solomon Islands, who likely used the stars and wind to orient, as well as wave piloting to anticipate land by detecting how islands disrupt ocean swells and current. Eventually, the long-distance voyages that were necessary for trading food, animals, and information with other island communities in Micronesia decreased, and wave piloting was primarily used for interisland voyages between the Marshalls' two chains of islands and coral atolls that stretch over some one hundred thousand square miles. During this time, the practice of wave piloting reached the pinnacle of mastery. Polynesians use a combination of wind and stars to wayfind, and other Micronesian communities primarily rely on stars. But the Marshallese rely on waves almost exclusively to deduce the direction of land, and they can do so over hundreds of miles.

The idea to bring anthropologists, physicists, and oceanographers to the Marshalls to study wave piloting was Korent Joel's, known as Captain Korent. From the atoll of Kwajalein, and one of the few remaining people on the islands who practiced traditional navigation, Joel hoped that Western scientists would not only validate the knowledge behind the practice but also increase its prestige and help him to preserve it for the next generation. His main desire, Genz told me, was to obtain a computer simulation of how the waves work that would both enhance his abilities and help teach subsequent generations. There was urgency to Joel's mission. For hundreds of years the Marshall Islands' indigenous navigation knowledge and practices had been eroded by colonialism. First the Germans, then the Japanese, and finally the Americans had brought social disruption in the form of new economies and technology, missionaries, and disease. The Germans and the Japanese banned voyaging altogether, considering it dangerous and threatening to their control and trading companies. By 1910, nearly all the Marshallese chiefs, called *Irooj*, were using European boats instead of the traditional canoes. It wasn't just navigation that was affected: other knowledge systems, from medicine to textiles, oral storytelling, chants, and songs, also disappeared.

Then, after all of that, the people of the Marshall Islands underwent what Genz has described as one of the most violent histories of the twentieth century. The Pacific War brought air raids and hunger, and when the war was over the American military began twelve years of nuclear weapons testing on the atolls of Bikini and Enewetak at the northern end of the island chain. During those years, sixty-seven atomic and thermonuclear bombs were detonated; one of them, called "Castle Bravo," created a blast a thousand times the magnitude of the Hiroshima and Nagasaki blasts. When Castle Bravo exploded on March 1, 1954, it completely obliterated three nearby islands, and the atoll of Rongelap some hundred miles away, which had not been evacuated, was hit with the radiation fallout. For a day and a night its islanders were blanketed in snowlike ash and began experiencing radiation sickness, including severe burns. After a few days, they were finally evacuated, then returned in 1957 despite the continued risks. Some three hundred people, many who had lived through the blast and were suffering from thyroid cancer, self-evacuated from Rongelap in 1985, permanently leaving their home behind.

Before the nightmare of nuclear testing, Rongelap had been the site of the Marshall Islands' only navigation school. Apprentices from nearby atolls went there to receive formal training because its circular coral reef gave elders a real-life teaching model of how atolls affect the flow of swells and currents. Students would start by learning from stick charts, then by floating in canoes and feeling waves in the coral reef. Eventually they would take the tests necessary to earn the title of navigator according to long-standing Marshallese tradition, which carefully controlled the transmission and inheritance of such knowledge. During their final test, in which they apply their learning during a days-long voyage to a specific atoll, the student navigators experience what is called *ruprup jokur,* which means "breaking open the turtle shell," a kind of intellectual process in which their minds fill with knowledge. All of this was lost after Castle Bravo. "The physical and social consequences of the massive radiation fallout from the 1954 Bravo test on Rongelap, Rongerik, and Ailinginae atolls essentially terminated the transmission of navigational knowledge to a young generation of navigation students," wrote Joe Genz.

Korent Joel was one of those navigation students. As he told Genz,

> When I saw the bomb's light, I stood up in the house and looked down, and it was very bright. I didn't know what it was but I thought it was a very big moon. Wow, all very bright, I could see some smoke, but I was thinking at the time it was a cloud. A very big cloud. It's only 120 miles from Kwajalein to Bikini. So I can see the cloud and it's really a big cloud. The women there, my mother stayed on Ebeye but all the children of my mother's younger sister, they all got burned, even my grandma and grandpa they all got burned, all of them there. . . . All the elder navigation teachers were from Rongelap. Some of them died, some of the stayed. . . . Of all the elders, half died. They were old, very old. There were some that were sixty years old, some seventy years old. I saw those guys. They bathed in water that was contaminated. The Rongelap people have stopped going to the place to learn navigation because there are no more people there. . . . I would have learned many things if I had stayed on the island, such as making canoes.

Although Joel had never received the title of navigator after the loss of the school, he continued to learn skills directly from his grandfather and was among the handful of individuals who managed to practice traditional navigation in the decades that followed. When the last of the titled navigators died in 2003, he realized he needed to find a way to teach a new generation.

Huth and the party arrived in the capital of Majuro and were met by Alson Kelen, originally from Bikini Atoll and a student of Joel's. Kelen is also the director of Waan Aelon in Majel, meaning Canoes of the Marshall Islands. Known as WAM, the vocational skills school has emerged as the Marshalls' last, best hope for passing on navigation skills to the islands' young people. Over the years, Kelen and Joel successfully negotiated with different *Irooj* in order to both respect traditions that strictly prohibited the dissemination of navigation knowledge outside of family lines and begin teaching traditional canoe-building and navigation to at-risk youth. Both made the preservation and dissemination of Marshallese navigation knowledge a mission, traveling throughout Oceania to collaborate with boatbuilding and navigation schools, soliciting money for WAM, and communicating with academics and scientists. (Early in the revival project they enlisted Ben Finney, the American anthropologist who helped launch Hawaii's own resurgence in traditional navigation in the 1970s by cofounding the Polynesian Voyaging Society.) Kelen describes the canoe as the medium through which they teach young people at WAM their entire culture—carving, language, songs, stick charts, and navigation. "We started canoe races, going to all the different mayors of the atoll and asking them to bring the best of the best," he said. "No one wants to lose and so canoe knowledge has been expanded."

At sunset on June 22, Huth and the expedition set out from Majuro on two different boats—a traditional outrigger canoe called a *walap* and a chase boat. Their destination was Aur, an island sixty miles north, where they would deliver supplies and spend a few days before making the return journey. For the wave pilot, navigating at night is easier than during the day, when one's eyes can play tricks on you. In order to sense the waves, you don't look but rather feel them in your stomach as you lie on the boat. Kelen describes wave piloting as both a "feeling and the image that is in my mind. I picture where I am going," he said. "We are feeling our way with currents and swells. And we always joke around that's why Marshallese have big stomachs—we navigate with our stomach." But as they set out for Aur, the conditions proved difficult. The waters were choppy, a mess of heaving and contradicting swells and winds. Huth and others became seasick. Even Kelen found it hard to eliminate information in order to identify the strongest swells and deduce his direction. Still he managed to accurately calculate when they were fifteen miles off the southeast corner of Aur, making it possible for them to successfully cut into the extremely small opening in the reef, a dangerous maneuver.

The return trip was easier, and the scientists were able to gather more data. The trade wind swells always seemed to be coming from the east, and Kelen detected a current sweeping them to the west. During the night Huth lay in the middle of the back of the canoe and focused on the motion of the boat, sensing a rolling that came in threes and then a pitch to the north. He made a note of it, thinking: I'm staking myself to this observation. Huth wondered whether he may have detected what is called *dilep*. Meaning "backbone," it refers to a mysterious path of wave patterns that create a line between two islands. Marshallese navigators can follow *dilep* from one atoll to the next by detecting with their bodies the signature wave patterns emitted from the destination island. *Dilep*, as Genz points out, is the navigator's highest art, but "from our scientific perspective, we cannot explain why this succession of distinctive waves forms on the direct sailing course between islands rather than on either side of it."

Huth told me that during the three weeks he spent in the Marshalls, much of it was spent in his hotel room scouring maps and crunching the data generated by the trip to Aur. He realized that most travel takes place along the two chains of islands, Ratick and Rallick, which are oriented on a southeast to northwest axis. "I think the trade winds might hit these chains and give you a unique fingerprint, and *dilep* might be the path between atolls that you can follow perpendicular to the swells and waves created by the wind," he said. It occurred to Huth that the size of the canoe tuned out the shorter frequency information from these waves, and by lying at the bottom of the boat, sailors can gather information from feeling the waves and follow their paths over long distances. *Dilep* might have even less to do with the actual swells and more to do with the distinct motion of the vessel riding them. It's a hypothesis, one that he intends to test again.

Two years after the trip, I went to Harvard again to see Kelen, Genz, and van Vledder convene to discuss a planned scientific paper publishing their intial findings. I met Kelen, tattoos covering his arms and neck, and talked about his students at WAM, some of whom were about to graduate with certificates in advanced carpentry and entrepreneurship. For him, this meeting was just another stop on a seemingly never-ending travel schedule across the South Pacific and beyond, connecting people and information in the hopes of preserving Marshallese wayfinding and contributing to the preservation of navigation across Oceania. Van Vledder brought with him a computer model that reconstructed the conditions from their voyage to Aur. It showed that winds and swells had indeed come from the east—but also that there was a swell from the north, thereby confirming Huth's observation and perhaps bringing the group a step closer to explaining *dilep* and its significance for oceanography. As van Vledder put it, "Although we use computer models, our knowledge is incomplete. We try to reconstruct it and they are all simplifications. We can learn a lot from what the navigators know."

The collaboration between the scientists and the navigators was meant to benefit future generations of Marshallese navigators as well. Six months earlier, Joel had passed away from

complications of diabetes and an infection. As the lineages of traditional wayfinders faded, perhaps younger Marshallese could learn
from a body of science explaining traditional navigation practices.
But the collaboration presented a risk too. In the history of contact
between the Marshallese and modernity, the islanders had already
lost so much—language, traditions, physical health, and homes.
What might they lose in this latest iteration of contact, no matter
how benevolent its purpose? This concerned Huth. If they figured
out the scientific explanation of wave piloting, such knowledge
could be taught in the future, but in a very different way than how
it had been traditionally learned through rigorous training, experience, and the cultivation of fine senses. "When I asked Alson
how he knows there is a current, he just says he feels it," said Huth.
"So should we teach the Marshallese the science? We have to be
really careful." But Joel would likely have disagreed, Genz told
me. It was Joel who first invited scientists to build a computer
simulation of the waves, and who had nourished a vision in which
Marshall Islanders married indigenous and scientific ideas for
the betterment of their community. This wasn't a concession but
an act of agency, selecting and utilizing technology and information in pursuit of a goal: to ensure that future minds continue to
undergo *ruprup jokur* and fill with knowledge of the sea.

ASTRONAUTS OF OCEANIA

Off the southern tip of Manhattan there is an island where the Lenape Indians once used to gather nuts from chestnut and oak trees. Today it is called Governor's Island, and on a summer day I pushed my two-year-old son onto a ferry going there to join hundreds of Hawaiians as they witnessed the departure of *Hōkūle'a*, a double-hulled Polynesian voyaging canoe that was at the tail end of a trip of some forty-seven thousand nautical miles, eighty-five ports, and twenty-six nations, all navigated without the use of Western instruments, maps, or charts. The purpose of the round-the-world journey was to spread awareness of *malama honua,* a Hawaiian concept of caring for the earth, at a time of climate upheaval for the world's people, particularly those in the South Pacific. The Polynesian Voyaging Society, which built the canoe in the early 1970s, described the trip as creating a lei of stories, big and small, to bring people together.

At a conference earlier in the week I had heard the *Hōkūle'a*'s mission described in much more radical terms, as a global movement for the resurgence of indigenous knowledge, language, and land-based practices that together represent an alternative future for humanity. Oceania has produced some of the world's richest, most complex, and beautiful traditions of navigation, many of which were lost after first contact with Europeans. Like in the Marshall Islands, colonial governments had at times banned interisland travel and even forced the use of navigational instruments on their indigenous subjects. The British did it in Kiribati and Fiji. The French did it in Tahiti and the Marquesas, where they made it illegal to sail without a compass. In Hawaii, traditional canoes and navigation practices faded completely for several hundred years. However, some, like the Caroline Islands, managed to maintain their navigation traditions through the twentieth century. Today traditional navigation is a nexus for cultural empowerment. Nā'ālehu Anthony, a filmmaker and crew member of the *Hōkūle'a*, argues that the teachings of Oceanian ancestors have become effectively guidelines for indigenous resurgence and determination. "If you want to cause change, disruption, resurgence, you can look at what they do on the canoe. They are practicing it every day. Some people say the navigator makes six thousand decisions a day. What direction, speed, how many miles? Open the sail, close the sail? Is a crew member sick? Those six thousand decisions they make cause change, and positive change in the right direction. If you want to cause disruption and change the arc of where everything is going, you can do so with decisions you make every day." Anthony called Oceania's original voyagers "the astronauts of our ancestors. They were the explorers of the earth. They mastered the science of living sustainably on islands. We're resurrecting indigenous wisdom and sailing it in the way of the ancestors. We've reawakened a cultural pride, identity and connection to place."

Vicente Diaz, a professor of American Indian Studies at the University of Minnesota and a traditional navigator from Guam, also described the building and sailing of traditional canoes like *Hōkūle'a* across Oceania as part of a crucial effort to decolonize Oceania and as symbolic of indigenous struggles for self-determination. But such struggles, he warned, can and will be co-opted by the status quo. "The welcoming of the *Hōkūle'a* in New York City should be a cause for concern," he told the audience.

"The radical potentials of traditional voyaging practices are worth celebration, but we can't shy away from the political struggles that arise at every turn in indigenous revitalizations." In other words, the *Hōkūleʻa* is a power grab by the disenfranchised, and one that threatens the continued legacy and policies of colonization across the South Pacific.

Diaz, speaking in 2016, pointed out that one of James Cameron's upcoming sequels to *Avatar*—a movie that exploited the cliche of a White Messiah saving Natives—was inspired by Cameron's deep dive into the Mariana Trench in the western Pacific Ocean. Cameron is just re-creating the same old "neoimperialist dream of conquering," he said. Meanwhile, he pointed out, the Disney Corporation was just a year away from releasing *Moana*, a film set two thousand years ago in Oceania, about a sixteen-year-old girl who sets out on a quest that requires her to learn the ways of traditional sailing and navigating. "I dread it coming out," said Diaz. "Behind it there is a marketing machine that is so large and so precise that it *will* become the dominant narrative." The fact that Disney would be presenting Pacific spirituality and culture to a global audience, and would (and did) make millions of dollars from the enterprise, galled him. The film's romantic and mysticized explanations of indigenous wayfinding erase what Diaz sees as the cultural specificity and historicity of Pacific seafaring's "science and technology."

Hōkūleʻa had in fact been built to counteract false narratives about the South Pacific, specifically how the original colonization of Oceania had taken place. The books *Kon Tiki* and *The Ancient Voyages of the Pacific* purported that South Pacific Islanders could not have voyaged intentionally across the ocean because they lacked the material technologies of the West. It really wasn't until the 1960s that scholars began to document and study South Pacific wayfinding traditions and realized they were still practiced in places like the Marshall and Caroline Islands. One was Ben Finney and the other was David Lewis, the New Zealand doctor who would later go to Australia to try and understand the skilled navigation practiced by Aboriginal people in the Western Desert. Finney and Lewis believed that the original colonization of Oceania couldn't have been accidental. They found support for their ideas in the work of anthropologist Thomas Gladwin, a former British administrator who was posted in the Caroline Islands in Micronesia. In 1967 Gladwin returned to an atoll called Pulawat to take a course in noninstrumental wayfinding. He was particularly interested in "the process of thinking" among people who were completely outside of a European system of practical knowledge. Perhaps the ability to find, record, and analyze a coherent body of "native" knowledge, he thought, would provide new insights into human cognitive processes, and even the perceived differences in intelligence between lower and upper classes in Western society.

In Pulawat, Gladwin found a culture in which voyaging was at the heart of the very purpose and meaning of life. Their suspicion of mechanization and motorboats was high, and "almost every young man seems still to aspire to become a navigator," he wrote. Children were taken on their first trips at the age of four or five, and sailors took tremendous joy in launching challenging, impromptu journeys on the numerous twenty-six-foot sailing canoes on the atoll. "The voyages are often organized on the spur of the moment, and an expedition to Pikelot even grows occasionally out of a long drinking party. 'I'm going to Pikelot. Who's going with me?'" Gladwin described. In one sixteen-month period, he recorded seventy-three interisland voyages. One of the most common was 130 miles to the atoll of Satawal, and in 1970, a 450-mile journey was "reopened" after seventy years of disuse when a captain received orally transmitted directions to Saipan and navigated it. "With such abounding enthusiasm for the sea it is evident that taking a trip to another island becomes in large measure an end in itself," Gladwin wrote in his book *East Is a Big Bird*. Even as the Pulawatans had converted to Christianity and passenger boats began serving the islands, Gladwin observed that the islanders' commitment and passion to navigation seemed to grow stronger rather than weaker. Pulawatans felt such élan for canoe trips that it wasn't uncommon for them to launch one to Truk, 150 miles away, just to get a preferred type of tobacco.

———

Gladwin made the analogy that just as Westerners who drive automobiles often see their home as a location on a larger road map, Pulawatans see their islands as communities connected by lines of travel through the ocean. They maintain vast mental maps of all the islands' spatial relationships to one another, but because these places are all out of sight of each other, the "landmarks" they use are swells, animals, reefs, wind, the sun, and, most important, stars. In some places, these cues were described in song. Vicente Diaz has described learning from the late Sostenis Emwalu of Pulawat about chants that are lists of creatures, stars, reefs, and landmarks referring to the voyage from the Central Carolines to Saipan through Guam. Diaz says, "[W]hen set to tune and performed properly this list was nothing less than an ancient and time-honored mnemonic map for travel." From a young age, Carolinian navigators memorize star "courses" or "paths," the points on the horizon where sequences of stars rise or set over an island, in order to create routes. In many cases, the navigator knows star paths for islands he has never even been to before. All together, Gladwin estimated, Carolinian navigators commit to memory star courses for over a hundred islands that span several thousand miles, and to sail accurately from one to the other they use a system called *etak*. "This is a body of knowledge which is not kept secret," he wrote, "but there is scarcely any need to do so. No one could possibly learn it except through the most painstaking and lengthy instruction, so no outsider could pick it up merely by occasional eavesdropping."

With *etak,* the navigator chooses an island, real or imagined, along his route as a reference point. He then uses the star bearings for this island to judge the distance he has traveled; each passing star overhead makes up a sort of segment of travel, and the number of segments along the voyage is the number of *etak*. Gladwin pointed out that the *etak* system isn't based on environmental input but on conceptualization—it's a tool for the navigator to synthesize his knowledge of rate, time, geography, and astronomy and create a framework for his powers of dead reckoning, the mental estimate of course and distance by sensing speed over time. "Everything that really matters in the whole process goes on in his head or through his senses. All he can actually see or feel is the travel of the canoe through the water, the direction of the wind, and the direction of the stars. Everything else depends upon a cognitive map, a map which is both literally geographical and also logical," wrote Gladwin.

Understanding *etak* is tricky. Pulawatans don't think of themselves as moving toward a star, but rather imagine the canoe and the stars are staying still while everything *else* is moving—water, islands, wind. It's an experience, Gladwin explained, akin to a passenger on a train looking out of a window as everything passes by.

> This picture he uses of the world around him is real and complete. All the islands which he knows are in it, and all the stars, especially the navigation stars and the places of their rising and setting. Because the latter are fixed, in his picture the islands move past the star positions, under them and backward relative to the canoe as it sails along. The navigator cannot see the islands but he has learned where they are and how to keep their locations and relations in his mind. Ask him where an island is and he will point to it at once, probably with considerable accuracy.

While Gladwin's research focused on the depth of knowledge of one specific atoll in Micronesia, David Lewis's book *We, the Navigators* recognized a commonality among the South Pacific navigation traditions that were still practiced. He saw a shared Pacific Island system of which the individual island skills were aspects, whether it was *etak* in Pulawat or wave piloting in Rongelap. In 1973, Finney, artist and historian Herb Kāne, and Hawaiian surfer and canoe paddler Tommy Holmes decided to build a canoe and sail it to Tahiti using traditional navigation as a way of proving that South Pacific Islanders had used these skills to voyage across Oceania. The effort became a catalyst for the emerging Hawaiian Renaissance: a resurgence of pride in indigenous music, art, farming, and sport. As the author Sam Low writes in his history of the *Hōkūleʻa*, *Hawaiki Rising*, Kāne believed that "the canoe was the center of the old culture—the heart of a culture that was still beating—and I thought that if we could rebuild that central artifact, bring it back to life and put it to hard use, this would send out ripples of energy and reawaken a lot of related cultural components around it."

They set out to raise $100,000 and started enlisting others to help design the canoe. When the boat was launched into the water a couple of years later, they brought a man named Pius Mau Piailug to Hawaii from Micronesia. He had been born in 1932, and both his grandfather and father were master navigators. The place of his birth was Satawal, the atoll 130 miles from Pulawat where Gladwin had spent months learning navigation from the Caroline Islanders. By eighteen, Piailug had been initiated in the *pwo* ceremony, a sacred initiation for navigators. But after that, the ceremony stopped being held for lack of students, and Piailug worried that the knowledge was going to be lost to the next generation of Satawalese. He became friends with a Peace Corps volunteer by the name of Mike McCoy, who told Finney about Piailug. The connection was extremely fortuitous; Piailug possessed a body of knowledge that had disappeared in Hawaii but that a group of young people were now eager to revive, in particular Nainoa Thompson, a young canoe paddler who began years of study under Piailug's instruction but also utilized resources at a local planetarium, developing a hybrid of techniques from both old and modern sources.

The first successful voyage of the *Hōkūleʻa* to Tahiti took place in 1976 with Piailug as navigator. Then in 1980, after years of study, Thompson navigated the canoe to Tahiti and back. Thompson describes a moment during that voyage when he nearly lost his way, only to glean what he understood to be the mystery of wayfinding in the ocean. One night he lost track of all the directional cues in bad weather and began to panic. He experienced an overwhelming sense of being out of control, of the dread of failure, when suddenly he *felt* the moon above him. Based on this single sensation, he was convinced he knew where he was and managed to navigate the canoe. "I can't explain it. There was a connection between something in my abilities and my senses that went beyond the analytical, beyond seeing with my own eyes," he said in *Hawaiki Rising*. "That night, I learned there are levels of navigation that are realms of the spirit. Hawaiians call it *na'au*—knowing through your instincts, your feelings, rather than your mind or your intellect. It's like the doors of knowledge open and you learn something new. But before the doors open you don't even know that such knowledge exists." In 2007, Thompson and four other Hawaiians, as well as eleven others, were initiated in the first *pwo* ceremony on Satawal since 1950.

———

Few predicted the *Hōkūle'a* would still be sailing forty years later. But during that time South Pacific Islanders have undertaken a fight to preserve navigation practices, or in some cases resurrect them entirely, in tandem with a broader cultural renaissance in language, art, and education. They've managed to do this by turning to elders, sharing knowledge between islands, and launching educational organizations, canoeing clubs, and schools. The *Hōkūle'a* helped stave off a cultural extinction, and today, in addition to the Polynesian Voyaging Society, Hawaii now has over a dozen voyaging societies, and there are also voyaging societies in the Cook Islands, New Zealand, Fiji, Samoa, Tahiti, and Tonga. There are indigenous skills and navigation schools like WAM in the Marshall Islands and the Waa'gey school in the Caroline Islands. Some of these resurgences are taking place in less expected places: a San Diego community of Chamorros, the indigenous people of the Mariana Islands, spent a year building a traditional forty-seven-foot canoe from a single redwood tree—the first to sail in nearly three centuries.

In these places, practicing navigation traditions is an act of self-determination and authority, taking control over identity by wrenching it away from missionaries, colonial governments, and tourism economies—even scientists and anthropologists. On the leeward side of the island of Maui I talked to Kala Tanaka Baybayan, a thirty-three-year-old mother of two and native Hawaiian, about this effort. We met in Lahaina, a laid-back town whose main street is lined with upscale surfing boutiques and tourist bars selling happy-hour mai tais, and greeted one another in an oceanfront park under a *halau*, a wooden pavilion with a traditional fiber roof that serves as a "house of instruction." The ground beneath our feet was once "crown land," the royal grounds for Maui's chiefs and kings going all the way back to the the sixteenth century's chief Pi'ilani. After Kamehameha the Great took control of the Hawaiian Islands in the early 1800s with a fleet of 960 war canoes and some 10,000 soldiers, Lahaina became the capital of the kingdom. Looking west toward the ocean, I could see the reef offshore where members of the royal family once rode the longest continuous wave on Maui on their *papa he'e nalu*, surfboards. Baybayan told to me that a lizardlike deity, called a *mo'o*, has always protected the people here; when some of Maui's residents decided to build a traditional sea-voyaging double-hulled canoe in the mid-1970s, the first in many generations, they called it *Mo'olele*, the leaping lizard. The boat, called a *wa'a*, is a forty-two-foot canoe with solid wood hulls and was dry-docked next to us.

Baybayan describes herself as part of the new generation of Oceanic voyagers. "We're using tradition and knowledge but also science and new heuristic devices," she says. I noticed an unusual tattoo on her left forearm, a pattern of abstract shapes in black ink. She explained to me that they are symbols for navigation done in a Marquesian Island style. "The T shapes represent an octopus, Kanaloa, the God of the Ocean whose tentacles stick onto knowledge. The triangles represent the stars, and the birds are the ones that help clue us to where land is," she said. "I got it in the beginning. When I decided I wanted to do this forever." Baybayan has studied traditional wayfinding for twelve years as an apprentice navigator; in two months she would board the *Hikianalia*, the seventy-two-foot sister canoe to the *Hōkūleʻa*, as its captain, navigating from Hawaii to Tahiti to meet the canoe and celebrate the end of its four-year-long voyage around the world. When she's not at sea, she teaches Maui's schoolchildren about Hawaiian voyaging and wayfinding as the education coordinator for Hui O Waʻa Kaulua, the nonprofit that originally built *Moʻolele*.

Baybayan's maternal great-grandparents were from Hiroshima, and her father's family were fishermen from Lahaina. In the late 1970s her dad, Chad Kalepa Baybayan, became interested in traditional voyaging and went to learn from Nainoa Thompson. "I could see that Nainoa was always looking for signs—the swells, the stars, the wind—and I tried to see what he was seeing," he told Sam Low in the book *Hawaiki Rising*. "I knew that he was in a different world from the rest of us and I was curious about what kind of world that might be. Just watching him, I began to dream that someday I could be a navigator. I knew it was a far-fetched dream but that was my opportunity so I decided to learn as much as I could." He became one of the early crew members of the *Hōkūleʻa*, and in the spring of 1980 he was one of fourteen who sailed the canoe thirty-one days from Hawaii to Tahiti, a distance of twenty-four hundred miles, using only the stars, wind, water, and birds. When the tops of the coconut trees of an island in the Tuamotus became visible on the horizon, he described the experience as time running backward, of reliving the voyages of his ancestors. "My *ʻaumakua* were with me," he said. Today he is navigator-in-residence at the 'Imiloa Astronomy Center of Hawaii.

But Baybayan wasn't ushered into voyaging by her father. She describes him as a quiet person whose voyaging often took him away from home. Though he was fluent in Hawaiian, she never learned the language from him. "He never asked us to go sailing when we were young. It was his passion, and he wanted us to decide what our passions were," she explained. When Baybayan went to college at the University of Hawaii in Hilo and Maui, she developed a strong interest in traditional Hawaiian culture. She became fluent in Hawaiian and learned the history of her birthplace, how the islands were inhabited by long-distance voyagers. "In my twenties I wanted to know about voyaging. I asked my grandma. She said I'll go talk to your dad, and then she [came back and] said you're going to sail with your dad. He has just finished building a voyaging canoe."

Her first voyage was from Oahu to Lahaina, a journey that took one day and night. "Before that I was into typical, regular things. I would say into pop culture. This was my first experience into another world. I realized that there is so much more, there is so much more to what we do in our everyday lives. If you're paying attention and listen, often there is a whole other story, telling us where we are." Baybayan kept learning, availing herself to crew on any voyages she could. "In the beginning I didn't know what to ask. The more I did it, the questions would come to me and it would take me to the next level. I could draw the connections. I could follow the path that the story follows," she said. "It's a gate to other sets of questions." In 2007 she sailed to Japan on the *Hōkūle'a*, the same year that her dad and four other men were initiated in a *pwo* ceremony and became a part of the two-thousand-year-old lineage of master navigators, by Mau Piailug himself. In 2014 she was a crew member on the *Hōkūle'a* from Hawaii to Tahiti. "When I'm away from land and cut off and on the canoe, my mind changes. Immediately, in the first two days, I'm different. I'm seeing things," she said. "It's 100 percent relying on yourself to see, hear, and feel, there's no modern instruments, there's no compass. It makes me feel good. I'm not lost." Fear, she said, never enters into the picture. "I'm more nervous during the last ten minutes of the voyage, when we come back to harbor."

For Baybayan, navigation is a cultivation of both scientific knowledge—geometry, physics, mathematics—and intangible instincts, a set of intuitions that are nourished and strengthened through experience and eventually, over a lifetime, add up to mastery. "Stars are actually the easiest part of the puzzle," she said of what's happening in her mind when she navigates. "We do more dead reckoning. I can't overcomplicate things. Was that seven knots or six knots? You have to be confident. All of the training is to build that confidence in us to make these observations." There are times, she explained, that utilizing navigation skills feels more akin to spirituality than science. "Some people say traditional navigators were scientists but spirituality was a part of what they did. And a huge part of what we do is spiritual. Science doesn't account for spirituality." Her words reminded me of what Nainoa Thompson had said about a moment he lost his bearings on the 1980 voyage to Tahiti aboard the *Hōkūle'a*, about the mysterious "levels of navigation that are realms of the spirit."

Despite her twelve years of experience and captaining the *Hikianalia* through the deep sea over twenty-four hundred miles, Baybayan makes a living at home on Maui as a server at the Feast at Lele, a nightly banquet where tourists are treated to traditional hula dancing depicting the history of Polynesian migration and eat Kalua pig and *poke 'ahi*. On some nights she performs at the Westin Ka'anapali Ocean Resort, offering guests a chance to learn "Stellar Navigation." Few people, it seems, have found a way to survive and support a family voyaging canoes in the modern world. "If the voyaging societies were successful, then we would have voyaging canoes out there," she said, nodding her head toward the empty stretch of ocean behind us. "Every day we would be getting a different group of kids." She pauses. "It's scary, the fragility of this knowledge, because if you don't take care of it, you lose it. We're fighting hard."

———

Five thousand miles from Maui, I sat on a sloping grass field on the southern side of Governor's Island eating fried chicken, *musubi*, kimchi, and chunks of sweet pineapple, watching men and women dancing traditional hula while we all waited for the *Hōkūleʻa*. As word spread among us that the canoe would soon be swinging by the island, everyone moved to the waterside, the lower Manhattan skyline gleaming in the sun, and began to chant a blessing, wishing the crew of the *Hōkūleʻa* safety on the next leg of their journey. Then the giant canoe came into view, its two red crab claw–like sails pregnant with the wind, cutting across the choppy channel with incredible power and speed. As it passed by us, a crew member stood on the bow, puffed his chest, and blew into a conch shell, and the sound was carried by the wind across the water to us.

The dissonance between the proud boat sailing in front of the steel and glass infrastructure of the world's most powerful economic center was startling, and more beautiful than I imagined. I thought about something Anthony had said: for Pacific Islanders, the canoe itself is an island in the ocean, and that island is precious, a means of survival, one that makes us human. And the canoe-as-an-island is a metaphor that can be extended to a much greater scale, for what is the earth if not a kind of canoe upon which we are all floating in a sea of space, bound together by a shared destiny? He said, "What if it's not just Hawaii, not just Polynesia, not just the Pacific, what if that perspective was common to people all around the planet?"

Then the *Hōkūleʻa*, simultaneously a representation of an ancient past, a symbol of resistance to the present, and the crucible for so many people's hope for the future, disappeared out of sight.

NAVIGATING CLIMATE CHANGE

When I set out to talk to traditional navigators in the Arctic, Australia, and Oceania, I had not anticipated how intricately the issue of climate change would be intertwined in these conversations. Again and again the indigenous communities I visited happened to be on the front lines of climate disruption, and only later did I fully appreciate that it was often because of their unique cultural practices, including oral transmission of information through generations and methodical observation of nature, that they were so keenly aware of these disruptions. In some cases, individuals are able to compare the changes wrought by climate change against several hundred years of collective experience because of the integrity of their oral traditions.

In the Arctic, where sea-ice conditions, weather, and temperature are increasingly unpredictable, I learned that for years older hunters—the same individuals who most often possessed a mastery of wayfinding skills—have reported strange environmental phenomena that they said had never been witnessed before. The Inuit filmmaker Zacharias Kunuk was often told by these elders that the sun was emerging after the long winter in different places in the sky than before, and the stars were often appearing in the wrong place. Initially, Kunuk thought it was a joke, but the elders insisted; the earth must have shifted on its axis, they told him. "I never paid any attention to it, but when I started making the documentary *Inuit Knowledge and Climate Change*, people from other communities started saying the same thing," he told me. "I don't think they were taken seriously. The Inuit don't have a PhD, they haven't gone to the university." Kunuk was curious enough that he and a film partner, the geographer Ian Mauro, began reaching out to scientists for an explanation, even writing to NASA, which rebuffed them. They finally found a scientist at the University of Manitoba who was an expert in atmospheric refraction, a kind of mirage caused by changes in air density that deflects light. The earth had not shifted on its axis, but it turns out that the appearance of celestial phenomena *had* changed in the Arctic, most likely as a result of temperature variations stemming from climate change. As it turns out, the Inuit already had a word for this sort of mirage, but they didn't connect it to what they saw in the sky: *qapirangajuq*. It means "spear strangely," and describes how when spearing fish, a hunter has to adjust for refraction in the water. "You start realizing they are right," said Kunuk.

The scientific group Arctic Council, composed of members from eight countries, now predicts that the entire Arctic will be ice free in the summer by 2040. According to Claudette Engblom-Bradley, a professor of education, the Yup'ik in Alaska begin observing and predicting the weather as young children, but environmental changes are now making forecasts extremely difficult. Similarly, in villages like Qaanaaq in Greenland, the changes have wrought havoc on where people can go and what environmental cues they can use to inform their movements. As one villager, Jens Danielson, told the *Washington Post*, "Earlier, the hunters, they can just look at the weather and see how it is going to be the next few days, so I can go out. But today, you can't do that anymore, because the change of the weather happens from day to day, or from hour to hour."

In the South Pacific, climate change is literally confusing the environmental cues used for navigation. Seasonal trade winds are weakening or blowing in inconsistent directions. Across the South Pacific, home to ten million people, sea level rise threatens livelihoods and entire islands. These changes in sea level have been happening at an average rate of 1.7 millimeters per year for most of the twentieth century, a rate that accelerated in the 1990s, according to the Intergovernmental Panel on Climate Change. For islands like Tuvalu or Vanuatu and hundreds of others, the threat is flooding and erosion, and potentially the submergence of atolls. Ninety percent of the population of the Maldives, Marshall Islands, and Tuvalu live on land, for instance, that is less than ten meters above sea level.

The potential costs of relocating entire island populations in the future are head-spinning. As Robert McLeman points out in his book *Climate and Human Migration*, there is no direct modern-day analog for the disappearance of land masses for entire nations as could happen in the South Pacific. Such refugees would join the twelve million currently stateless people in the world. "Tuvaluans will not cease to be citizens of Tuvalu because of [sea level rise] but Tuvalu itself may physically cease to be habitable, becoming a sort of modern Atlantis," he writes. "No international law or policy would give automatic shelter or protection to those made stateless by [sea level rise]. Although the popular media, nongovernment organizations, and some scholars use terms like 'environmental refugees' or 'climate change refugees,' such a category of persons simply doesn't exist under international law."

Over the course of my reporting, I began to connect another surprising aspect of the relationship between human navigation practices and global climate. At the start of the Industrial Age, humanity unleashed a revolution in transportation powered by accessing fossil fuels deep in the earth, propelling us to greater and greater speeds in cars, airplanes, ships, and rockets, vehicles unimaginable to our ancestors. In that same period, from the invention of the combustion engine to now, we have put so much heat-trapping carbon into the atmosphere that levels are higher than at any period in the last eight hundred thousand years. In other words, the same revolution in transportation that led to changes in the way we navigate helped create the problem of climate change, and now climate change will undoubtedly impact how and where humans move in the coming decades.

But are indigenous cultural practices and knowledge of navigation potentially critical tools for combating climate change? McLeman writes that pastoralist cultures from Central Asia to Lapland to Saharan Africa practice inherently migratory lifestyles. Among the Aboriginal Australians, Inuit, and First Nations of North America, mobility and migration are inherent components of culture practice and environmental stewardship. What might others learn from those who embraced mobility and migration as part of their identity and possess skills for travel and survival under their own power? The Columbia University professor Rafis Abazov has written that the modern world has much to learn from nomadic cultures, including attitudes toward the Other, because exploring otherness and learning from newcomers about the land beyond the horizon is essential to nomadism. At the very least, it seems that indigenous communities have much to offer the scientific community if only their traditions and methods for accumulating and synthesizing knowledge were seen as equally valid.

The Indigenous Peoples' Bio Cultural Climate Change Assessment Initiative argues that indigenous knowledge, experiences, wisdom, and perspectives are needed to develop evidence-based responses for adaptation. And across the South Pacific, the revival of traditional navigation is increasingly seen as a potent response to the threat of climate change, and the specific technologies and economies that unleashed it. The Marshall Islands is the first Pacific country to commit to reducing its transport emissions by nearly 27 percent by 2030. The Okeanos Foundation wants to create a new Pacific interisland transportation industry using a combination of traditional canoe technology, biofuels, and solar power to wean Oceanians away from a dependence on the very

fossil fuels that threaten to submerge their nations. Voyaging societies, NGOs, nonprofits, schools, and communities are recognizing that traditional knowledge and wayfinding could be powerful
aspects of a sustainable, fossil fuel–free future.

———

At the University of the South Pacific on the island of Viti Levu
in Fiji, I visited the offices of the Sustainable Sea Transport Research Programme and its director, Peter Nuttall. Dedicated to developing low-carbon shipping solutions, the program is applying
traditional Fijian sail technology to commercial global sea transportation in order to end the region's dependence on fuel. The sea
transport industry as a whole is the sixth-largest emitter of greenhouse gases in the world; creating a fleet of sail-powered catama-
rans, Nuttall thinks, could create a carbon-positive alternative
with potential to remake the economy in ways that would support local indigenous tradition and skill. "Current transport, especially domestic, is increasingly unsustainable and owes nothing
to the Pacific's rich indigenous, historic, and arguably sustainable legacy of vessel/sail technology development," he has written. In Nuttall's alternative vision, the solution to the threat of
climate change and rising sea levels in Oceania is Oceanic history itself. But carrying out this vision depends on resurrecting
the nearly extinguished tradition of canoe-building, which is why
I was there: to go to the last village in Fiji whose members still
know how to build, sail, and navigate *camakau,* the country's traditional boats.

The story of how Nuttall discovered this village is one of myth-
like serendipity. Born in New Zealand, Nuttall likes to describe
himself as a grumpy old Kiwi sailor. His kinship with Fijians came
from a deep admiration for their sea-transport traditions and love
of the ocean, which was

> the connection, the interface, the facilitator between people
> and god, people and environment and of culture to culture.

Sea-going vessels were the pinnacle of societal achievement. They were the ultimate line of defence. Their design and functionality was radically different from that produced from any continental paradigm, almost Zen-like in its approach to finding ultimate form in simplicity and from a minimal resource base. Terrestrial design and construction was not the primary role of craftsmen but what naval architects and shipwrights did in their downtime. Their vessels were the products of cultures where metals were not an available option, where swimming and walking were equally important, where survival at sea, more so than on land, was primary.

Nuttall knew that Fiji was once part of a complex political and trade network that encompassed most of central Oceania, made possible by the large fleets of Fijian-sourced sailing vessels. But for decades Nuttall simply couldn't locate any extant examples of their epic sailing past: no traditional canoes or anyone who knew how to build them. Few people seemed to venture beyond the sight of an island or reef anymore; outboard motors had replaced sails, and to save money and fuel, boats were driven in straight lines that cut across currents and winds, ignoring directional cues. Long-distance travel between islands occurred most often by ferry, and some islands received infrequent if any service. The only *drua* in existence was the *Ratu Finau*, a boat built in 1913 and currently housed in the Fiji Museum just a few miles from where Nuttall and I sat during our meeting. He once saw a small derelict example of a *drua* in a village in Kadavu, but no one knew much about it or its history. In 2006 he witnessed several *camakau* sailing in an arts festival, but he couldn't track the boats down—they virtually disappeared afterward. His research in libraries and museums was coming up mostly empty. "All indicators were that *drua* culture was now consigned to history and museum artifact," he wrote. It seemed to him that the living sailing traditions were virtually snuffed out and could

only be pieced together from a scattered and incomplete historical record and a few black-and-white photographs.

Then in 2009 something miraculous happened. One evening
Nuttall was anchored in Laucala Bay near the University of the
South Pacific campus when around twilight he saw the silhouette
of a *laca,* a traditional sail, on the horizon. He described the events
of that night: "My young sons and I leapt into our sailing dinghy
and sped to intercept it. As we met, my sons were plucked from
our dinghy and swept onto a large *camakau* with a ragged patchwork sail by a laughing and obviously Lauan crew. '*Mai, mai
lakomai*—come, *kaiwai,* we go to drink *kava,*' they taunted as they
raced ahead of me and disappeared into a muddy mangrove-
shrouded creek in what I had always assumed was an uninhab-
ited piece of city shoreline." That night Nuttall sat in the village
of Korova until dawn, talking for the first time to Fijians who not
only sailed *camakau* but were, he realized, the last people in Fiji
who knew how to build them. They bragged to him that they had
never owned outboard motors. It seemed a miracle that he had
found this village of sailors, and yet the fact that these represen-
tatives of ancient tradition inhabited such an anonymous spit of
land in the shadow of Fiji's capital city was undeniably tragic.

After meeting Nuttall's partner, Alison, and their two young
sons, and introducing my partner and two-year-old son, we
formed a traveling gang and walked to the end of the campus onto
a stretch of grass between a ditch and a main road as cars sped by
us. We followed the road for half a mile north as it hugged the
edge of Laucala Bay and then turned off onto a dirt path leading
into a shadowy tangle of mangroves. The first sign of the small
village located there was the outpouring of children who ran to
greet us, delighted by the new visitors and the unfamiliar toddler
willing to be indoctrinated into their group. As we ventured further along the path I saw half a dozen small sailing canoes pulled
up onto the shore, and beyond them a cluster of cement-block
homes. We greeted the women and men of the village with *ni sa
bula vinaka* ("warm hello") and were led to a large open-air

meeting space with a corrugated tin roof and mats spread on the ground. A large carved wooden bowl with short legs on the bottom was brought into the center of our circle, and one of the women began to prepare *kava*, pouring water into the bowl and soaking the ground root of the *Piper methysticum* plant to create a muddy-colored brew containing psychoactive properties renowned for creating feelings of contentment. A young girl filled a half coconut shell with the drink and walked to each of us sitting in the circle. Before accepting the cup from her, we each clapped once with cupped hands and then drank it down, handing it back and clapping again three times. This was the practice of *sevusevu*, the respectful offering of *kava* by a host to a guest. Integral to Fijian culture, *kava* is sometimes called *wai ni vanua*, meaning "blood of the land."

Woven nets hung around the space, blowing in the breeze, and the attention of the group turned to two men in their sixties sitting in the circle. They sat on upholstered seats taken from a car or minivan, and the atmosphere was one of a jubilant court, the elders on thrones and us looking to them as deferential subjects. Juijuia Bera and his older brother Semiti Cama are the last Fijians who know how to build the *camakau* and the *drua*. Of the two boats, the *drua* was the most astonishing in its size and performance: built of wood, grass, nuts, stone, bone, and sharkskin, and containing not an inch of metal, the *drua* had two asymmetrical hulls and measured a hundred feet. One *drua* could carry two hundred to three hundred people and travel at speeds of seventeen miles per hour. In times of peace they were used for diplomacy and transferring people and cargo; in war they were used as battleships in massive fleets that acted as rammers, blockade runners, and troop transports. One fleet of canoes carrying warriors was called a *bola*; in 1808 the trader William Lockerby was chased out of Swaddle Bay by 150 canoes, and in the mid-nineteenth century one observer witnessed two *bola* in Laucala Bay, just to the east of where I sat.

Such displays of power are a thing of the past. Most Fijians today have likely never seen a *drua,* save for one on the national fifty-cent coin. But historians still consider the *drua* to be the pinnacle of Oceanic technological design, a canoe that was "far superior to those of any other islanders in the Pacific." The last *drua* voyage in Fiji was undertaken by Bera and Semiti's own father, Simione Paki, in 1992, when he sailed from the Lau archipelago to Suva. A Methodist preacher, Paki raised sixteen children on the island of Moce, traveling with them by *drua* from young ages so they knew their way around the boat and learned to navigate. Of the whole Lau archipelago, which contains the best navigators in Fiji, the Moce islanders are considered the most accomplished.

More *kava* was prepared, and Bera and Semiti described how they learned to sail and navigate, using the rising or setting sun to mark the direction of east and west on the horizon and the appearance of the planets Venus, Mars, Jupiter, or Saturn in the night sky. "Swells, wind direction. These are the things that can help you in the sea," they told me. "In traditional navigation you have to see the sun set every day." While traveling from one island to another, a star might be followed that marks the direction of travel, until it sets and another star would be chosen, a system of star paths common to Oceania. They also used familiarity with the currents and wind patterns of the Lau archipelago. In 1989, Bera and Semiti's father decided they would leave Moce and sail two hundred miles to Suva, where he hoped to start a business offering cruises on *drua* to tourists and Fijians. He brought Bera and his brother Metuisela Buivakaloloma with him, establishing a village on a sliver of land vulnerable to typhoons and flooding that no one else wanted. In 1993 Buivakaloloma sailed back to Moce to bring back a *camakau* and was lost at sea; his boat washed up on an island a month later. In 2004, Paki also died at sea. The canoe-building practices back home started to become extinguished. "There's no water on the island, we have to travel even to get water. Canoes are the only transport that we knew," said Bera through a translator. "The way we learned is back in the island. I saw it and perfected it and brought it down here. Moce is where the builders come from.

In the eighties was when it dropped off. People started relying on outboards. Building canoes was not easy. We have to go out without roads and cut trees. Have to prepare . . . rope, and have to have money. We left the island and the last canoe that I built, after that it ended. There weren't canoes anymore."

Three years earlier, Bera and Semiti had been visited by several producers from the Walt Disney Company. John Musker and Ron Clements, directors of animated blockbusters like *The Little Mermaid, Aladdin,* and *Frozen,* were undertaking a research tour for *Moana.* They brought a sack of plastic toys for the village children and several hundred dollars, drank *kava,* and asked questions about Fijian boatbuilding, lore, and navigation. The villagers signed a contract with the producers' lawyers, believing that it entitled them to monthly payments in return for the information they gave Musker and Clements, money that would finally allow them to start the long-dreamed-of tourism business and build the first *drua* seen in Fiji in decades. But even as the first promotional illustrations for the movie were released, showing a Fijian *camakau* like those the producers had seen on the banks of the village in Suva, the money never came. In late November 2014 Nuttall met with the frustrated members of Korova village and afterward sent an email to Musker, reminding the producer that Nuttall had previously begged them not to rip off the heritage and indigenous knowledge of the village and their intellectual property rights. Nuttall stated the painful irony of the *Moana* endeavor for Korova's community members: "Seems to me Disney is about to launch a new virtual canoe across the horizon and that you will find that a cause of celebration and success and profit. The saddest outcome I can think of would be to find one of the Korova kids . . . playing with a broken plastic *camakau* from McDonald's while sitting on the last rotten *camakau* hull on a rising tide."

Indeed, Disney was already undertaking merchandising partnerships with LEGO, Subway, and other toy companies. The company released a Halloween costume for the film's character Maui: a brown-colored shirt, pants, and wig that turned wearers into a dark-skinned, tattoo-covered Polynesian demigod with long hair. The outrage over this "Polyface" costume was so swift that Disney made a public apology and took it off the shelves, assuring the public that the company had taken great care to respect the cultures of the Pacific Islands that inspired the film. When the movie was released in 2016, it grossed over $56 million in two days. Over the next nine months it grossed $638 million. Yet the community at Korova waited years for any compensation; eventually Disney made a donation to a trust managed by a foundation for the construction of a boat. During that period Nuttall's oldest son used a small inheritance he received after the death of his grandmother to fund the effort of building a *drua*.

I was curious: why had the community decided to share their knowledge with the Disney producers in the first place? Jim, Bera's nephew, was sitting next to me and explained, "Some people want to keep it secret. We don't love that. We believe that to live by the sea, you must not tell lies, and [in return] the ocean will protect you. It's against what we believe. All that's why the ocean is protecting us," he said. "It's the ocean that is the boss and love. The ocean is clean, powerful, friendly. The ocean can be dangerous if you go against it. That is what the elders, our father, taught us." Knowledge itself cannot be sold. "It is free to be given," Bera added.

The conversation turned from Disney to the Paris Climate Accords taking place in a couple of weeks. It was now night and my son was fast asleep in my lap; a few electric lightbulbs illuminated everyone's faces as they spoke about the meeting. A delegation representing indigenous communities would be in Paris, including a representative from Korova. Nuttall pointed to a young man in the circle who was visiting Korova to learn about canoe-building. He was from Tuvalu, a small island country north of Fiji that is extremely vulnerable to sea level rise, with some predicting it will become uninhabitable in the next century. "Even if the world stopped all the emissions tomorrow, Tuvalu will still be underwater," said Nuttall. "If he has to migrate, will he do it under his own sail? His ancestors could leave and go anywhere they wanted. But today we leave on a 747." He paused. "How do we maintain dignity in this crisis?"

The *kava* drenched my synapses, and the question seemed to hang in the air, essential and daunting. Perhaps it's hyperbolic to present navigation practices as a response to climate change; the practice itself can't reduce CO_2 emissions, and humans won't all turn in their cars or airplane tickets for sailboats. But, I thought, what if we took a more critical view of how we move through space? Considered how technology influences our choices and impacts the global environment? What if we paid more attention to the landscape around us, became witnesses to its patterns and changes, and shared that information with others? What if we nourished our attachment and concern for the places we live and travel to? *Those* things seemed to matter very much.

As the night drew on and the children disappeared to sleep, we gave our thanks and goodbyes, walked back to the car in the dark, and then drove on a winding highway along the southern end of Viti Levu. I sat in the backseat and my thoughts wandered. I was shaken by the tenuous survival of cultural practice and tradition, and the urgent reminder of anthropogenic climate change, the consequences of which were impossible to ignore in Fiji. But I also felt incredibly happy, warmed to the core by the hospitality and testimony of Bera and Samiti (and no doubt the free-flowing *kava*), who once again shared their thoughts and experiences with strangers. They wanted to give what they knew to others, and their dignity and vulnerability felt like a corrective to all the cynicism of the world. I looked out the window and saw the stars in the sky and remembered a passage in a book I had read by the neuroscientist Oliver Sacks. In the 1990s Sacks had traveled to the Micronesian atoll of Pingelap, some two thousand miles northeast from where we were now driving. He wanted to find out why the island had a disproportionate number of people with achromatopsia, the inability to see color. One night he partook in drinking *sakau*, as *kava* is called in that part of Micronesia, an experience Sacks recounts in his book *The Island of the Colorblind*:

> Knut, next to me, was looking upward as well, and pointed out the polestar, Vega, Arcturus, overhead. "These are the stars the Polynesians used," said Bob, "when they sailed in their proas across the firmament of space." A sense of their voyages, five thousand years of voyaging, rose up like a vision as he talked. I felt a sense of their history, all history, converging on us now, as we sat facing the ocean under the night sky. . . . Only then did I realize that we were all stoned; but sweetly, mildly, so that one felt, so to speak, more nearly oneself.

THIS IS YOUR BRAIN ON GPS

In the 1960s, the psychologist Julian Stanley became interested in understanding what makes child geniuses different from other children. What was the nature of their intelligence that made them so intellectually gifted? Stanley launched an investigation he called the "Study of Mathematically Precocious Youth," and half a century later, it has shown that the best way to raise a smart kid may be to nourish their ability to think spatially. This could mean engaging them in exercises that require them to imagine objects from different perspectives, or mentally manipulate images, and perceive patterns between them.

Over the decades, Stanley and his colleagues tracked the achievements of five thousand children who had scored unusually high on the SAT, some in the top 0.01 percent. From the start Stanley was interested in how the ability to understand and remember spatial relationships between objects might predict achievement and intelligence better than other tests such as verbal acuity. He regularly tested the study's participants on spatial aptitude, and, as the journal *Nature* reported in 2017, researchers decided to look at the scores on those tests and compare them to the number of patents and peer-reviewed journal articles the participants, many of them highly successful, had generated over the course of their careers. What they found was that the two data points were strongly correlated, so much so that David Lubinski, a director of the study, told a reporter, "I think [spatial ability] may be the largest unknown, untapped source of human potential." Raw intelligence, it seems, is intertwined with our brain's spatial cognition aptitude.

This insight has come at a time when young people in general are experiencing less and less demand to exercise their spatial navigation skills. As the neuroscientist Véronique Bohbot told me, she has begun to suspect that the sedentary, habitual, and technology-dependent conditions of modern living are changing how children and even adults use their brains. Bohbot, a researcher at the Douglas Mental Health University Institute and associate professor in the Department of Psychiatry at McGill University, has studied spatial cognition for two decades and believes that we are, in general, flexing our hippocampus less and less, with potentially damaging consequences. "People who have shrunk hippocampus are more at risk for PTSD, Alzheimer's, schizophrenia, and depression," she told me. "For a long time we thought the disease *causes* shrinkage in the hippocampus. But studies show that the shrunk hippocampus can be there before the disease."

Bohbot did her doctoral research with Lynn Nadel, coauthor of *The Hippocampus as a Cognitive Map.* "Back then the hippocampus was a fascinating brain structure to study, and it was the only structure known to be involved in spatial memory. But it was hypothesized that there were other brain structures involved in different ways to navigate in the environment," she said. In the mid-1990s at McGill University, where Bohbot was training rats in memory tasks, fellow researchers Norman White and Mark Packard discovered one other brain circuit was the caudate nucleus. Bohbot became fascinated by the implications of this discovery. Was it possible that people use very different brain structures for different strategies of navigation? If so, why? She began to conduct experiments in humans designed to distinguish which strategies were dependent on the hippocampus and which involved the caudate nucleus. What she found was that not only were the circuits different, but the strategies they corresponded to were hugely different too.

"The hippocampus is involved in spatial learning, i.e., learning to navigate using the relationships between landmarks," she explained. "Once you have learned the relationship between landmarks, you can derive a novel route to any destination from any starting position in the environment. Spatial memory is allocentric, it's independent of your starting position. You use spatial memory when you can picture the environment in your mind's eye. That's when you are using your internal map to find your way." Meanwhile, the caudate nucleus isn't involved in creating cognitive maps, it's a structure that builds habits. Using it, the brain can learn a series of directional cues such as "turn right at the corner with the grocery store" and "turn left at the tall white building," creating what are called stimulus-response memories. To understand what the caudate nucleus does, she told me to imagine how to get to the local bakery. "Every day you use the same route, and at some point it becomes automatic," Bohbot said. "You don't think about it anymore. You don't ask, where do I have to turn? Autopilot takes over. You see the white building, it acts as a stimulus and triggers a response to turn left to get to the bakery."

Although this strategy might seem similar to the egocentric strategy used in route navigation, it can actually be quite different. There are three types of stimulus-response strategies, according to Bohbot, and an egocentric strategy is just one of them. "An egocentric strategy involves a series of right and left turns that begin with your starting position: when you leave your home (the stimulus), you will turn right (the response). Then there's a beacon strategy where you could reach a target location from many different starting positions: the tall white building is your beacon (stimulus), and you head toward it, turning at every corner in its direction (response). And then there is the most common form of stimulus-response: a series of turns in response to various landmarks in the environment." Even though the caudate nucleus uses repetition to navigate successfully, it's actually *not* a spatial strategy. The key difference is that the response strategy doesn't involve learning the relationship between landmarks, so it becomes impossible to generate a novel trajectory in the environment. All the caudate does is signal—left or right—in response to a cue without engaging your active attention.

There is a persuasive evolutionary explanation for why nature invented this other (seemingly lazier) circuit: it means you don't have to retrieve a memory of a route or make spatial inferences every time you need to go home. It gives us the advantage of not needing to make calculations or decisions—or pay very much attention—to where we are going and how we are getting there. Autopilot is fast and efficient. "I don't have to think, that's great!" said Bohbot. But Bohbot also discovered a negative correlation between the two strategies: the human brain is using either the hippocampus or the caudate nucleus to get somewhere, but it never engages these two brain areas at the same time. This means that the more we use one, the less we use the other, and like a muscle that grows in strength and compensates for weaker muscles, a specific circuit can become preferred over time.

Scientists already knew that as we age, the strategies we use to move change. As children and young adults, we navigate and explore new spaces. Over time we increasingly rely on familiar routes and return to places that barely strain our cognition—we underuse our hippocampus. Each of our life histories likely traces this trajectory: we go from utilizing hippocampal spatial strategies to increased automatization. Bohbot discovered this when she undertook a study of 599 children and adults and compared the spatial strategies they preferred to solve tasks with. She and her coauthors found that children rely on hippocampal spatial strategies some 85 percent of the time, but adults over the age of sixty complete a virtual maze test using this strategy just under 40 percent of the time. The question remained, however, whether the preference for one strategy over another led to physiological differences in gray matter density and volume in the hippocampus.

In 2003 and 2007 in the *Journal of Neuroscience*, Bohbot and several researchers published two studies focused on measuring activity and gray matter in both the hippocampus and the caudate nucleus. They mimicked the classic spatial test for rats and applied it to humans by creating a radial maze in a virtual setting and asking participants to navigate it while they tracked their brain activity with fMRI. As expected, the individuals who used spatial memory strategies showed increased activity in the hippocampus, and those who used signal response had increased activity in the caudate nucleus. But then they went a step further and measured the morphological differences in these two brain regions for each individual. The researchers found a high probability that people who used a spatial strategy had more gray matter density in the hippocampus, and the inverse was also true: those who used a response strategy had more gray matter in the caudate nucleus. In and of themselves, these results might not be alarming. Successful navigators likely have an ability to deploy flexibility when it comes to these strategies: they can go on autopilot for speed and efficiency and engage cognitive mapping to solve new questions and challenges that they encounter. But what if we persistently prioritize the caudate nucleus over a hippocampal strategy? And what if this prioritization was not just occurring in some individuals in the population but was happening at a more endemic scale?

Bohbot told me she thinks it's possible that the conditions of modern life are leading us to flex the hippocampus less while spurring us to rely on the caudate nucleus. "Maybe in the past we never had to go on autopilot. Having jobs in one location and lives being more habitual is new. Industrialization learned to capitalize on the habit-memory-learning system," she said.

Compounding these societal changes is the fact that chronic stress, untreated depression, insomnia, and alcohol abuse can all shrink hippocampal volume. Anxiety alone has been shown to impact the spatial learning and memory of rats. Stress and depression seem to affect neurogenesis in the hippocampus, whereas exercise seems to improve learning and memory and resistance to depression, spurring a proliferation of new neurons. Patients with PTSD have been shown to have lower hippocampal volume, and one of the consequences of effective treatment for the disorder, such as the use of antidepressants and changes in environment, is increased hippocampal volume.

The widespread prevalence of these conditions has led Bohbot to be concerned that by the time children enter young adulthood, they might already have relatively shrunken hippocampal volume that makes them susceptible to cognitive and emotional impairments and behavioral issues. Indeed, an overreliance on stimulus-response navigation strategies does seem connected to a host of destructive yet seemingly unrelated behaviors. Because the circuit is located in the striatum, a brain area involved in addiction, Bohbot began to wonder: Would people who rely on a response strategy to navigate show any difference in substance abuse from those who relied on spatial strategies? In 2013 she published a study of fifty-five young adults that showed those who relied on response strategies in navigating had double the amount of lifetime alcohol consumption, as well as more use of cigarettes and marijuana. In a separate study of 255 children, she found that those with ADHD symptoms primarily rely on caudate nucleus stimulus strategies. More recently, Bohbot and her colleague Greg West showed that ninety hours of in-lab action video games will shrink the hippocampus of young adults who use their caudate nucleus, providing the first clear evidence that the activities we engage in can have a negative impact on the hippocampus.

Worst of all is the relationship between Alzheimer's and the hippocampus, which has been documented since the late 1980s. Hippocampal atrophy is associated with memory impairments in the elderly, and neuroimaging studies reveal that in patients with clinically diagnosed Alzheimer's, the presence of atrophy is nearly universal. Moreover, shrinkage of the hippocampus and neighboring entorhinal cortex predicts future diagnosis of Alzheimer's disease years later. This isn't surprising in light of the established links between hippocampal damage in amnesiac patients and their loss of spatial memory. Individuals with Alzheimer's undergo a painful process of losing both memory and identity. But one of the first symptoms is that they often lose their way, misplace things, and forget where they are and how they got there.

There may be genetic factors at work when it comes to the hippocampus and its relationship to Alzheimer's. As early as 1993, researchers had documented a risk gene for Alzheimer's called Apoipoprotein E or APOE. A year later, an allele of the gene (ApoE2) was found to be associated with *reduced* risk and delayed onset of Alzheimer's, slowing hippocampal atrophy. On the other hand, a different allele of the gene (ApoE4) indicated a *greater* risk for the disease. Young adults with the good allele seem to have more cortical thickness in the entorhinal cortex, which delivers inputs to the hippocampus, as well as a larger hippocampus itself. In her recent research, Bohbot has studied the cognitive correlates to the presence of these genetic traits in young adults. She genotyped 124 young people and tested them on a virtual radial maze: those with the good allele were more likely to use a hippocampal spatial strategy and possess more gray matter in that part of their brains.

Genetic predisposition might potentially limit atrophy, but does exercising spatial cognition prevent deterioration? Bohbot thinks that early interventions focused on spatial memory might actually decrease conversion rates to Alzheimer's, and that good spatial memory could protect individuals from the disease. Aging people who practice using their spatial memory have a more active hippocampus, a larger hippocampus, and better cognitive health. She has already found that participants who use spatial strategies show reduced risks of dementia when tested on the Montreal Cognitive Assessment, a test used to detect mild forms of cognitive impairment. Her work is now focused on finding ways to teach people how to improve spatial memory and their cognitive health. Among her recommendations are regular exercise, a Mediterranean diet full of omega-3 oils, meditation and deep breathing, as well as plenty of sleep. Most important, she advises actively building cognitive maps. Take new streets and shortcuts to get places; regularly draw a bird's-eye view of your environment with landmarks; incorporate new behaviors and routes into your daily life. The benefits of hippocampal health appear to be far-reaching. "There's some studies that show people who have a larger hippocampus have more sense of control over their lives," she told me. "What would that mean? One interpretation is that if you have better episodic memory, you can better remember what happened. And if you can better remember what happened, you can remember mistakes to avoid and good actions to repeat in order to obtain a desired outcome, and have a better sense of control. That itself is less stressful and you can better cope with things that happen in your life. Control is a mechanism to deal with adversity."

———

In the fall of 2017, Bohbot and ten other researchers published a report called "Global Determinants of Navigation Ability," in which they looked at the performance of 2.5 million people globally on a virtual spatial navigation task and then broke the data down to understand whether there were similar profiles in cognitive abilities between countries. One of the authors and architects of the study was Hugo Spiers, the neuroscientist at University College London who a decade earlier had studied the brains of London's taxi drivers, revealing that they possessed more gray matter in their hippocampi than bus drivers. At the annual conference of the Charles River Association for Memory at Boston University, Spiers presented the results of their findings from this latest study to an audience that included several eminent memory researchers, including Howard Eichenbaum. The data in Bohbot and Spiers's study was generated using a video game called *Sea*

Hero Quest. The game, which can be downloaded on any smartphone or tablet, is a spatial orienteering task in disguise. The goal is to navigate a boat in search of sea creatures in order to photograph them, and there are two ways to do this: players can travel along a twisting and turning waterway and then shoot a flare in the direction of the position where they started, or they can memorize a map beforehand that gives them a series of checkpoints they need to find their way to. The former is an example of dead reckoning (or path integration), while the latter is what the researchers define as wayfinding. Spiers reported that the game had been played three million times by people age eighteen to ninety-nine in 193 countries, from India to America, Brazil to Australia. The results were fascinating.

The data shows that spatial navigation ability starts declining in early adulthood, around nineteen years of age, and steadily slips in old age. People from rural areas were significantly better at the game. When it came to countries themselves, Australians, South Africans, and North Americans showed generally good spatial orientation skills, but the real outliers were Nordic countries. Players from Finland, Sweden, Norway, and Denmark, as well as Australia and New Zealand, showed the most accurate dead reckoning skills. What explains this? Spiers displayed a scatter diagram that showed a causation between GDP per capita and navigational ability. This might have something to do with factors such as healthcare, education, and wealth. But the truly indicative factor is not whether a county has a high GDP, but whether it participates in the competitive sport of orienteering, in which people use a map and compass to race each other to various checkpoints outdoors. It so happens that orienteering is hugely popular in Nordic countries. Spiers pointed out that the number of world championship medals won by Nordic countries between 1966 and 2016 is strongly correlative for how good players are at *Sea Hero Quest*.

Someone in the audience raised the possibility that the study data is skewed because only people who are confident at engaging with a virtual reality interface would voluntarily do the task. What I wondered was whether access to the internet or a computer had also confined the data. What if even the best in-game navigators of those three million people who have played the game were some of the poorest navigators on the spectrum of what is humanly possible? While players from northern European countries generally scored high on *Sea Hero Quest*, other studies hint that dead reckoning skills in these countries might be nothing special. For instance, the American linguist Eric Pederson has tested men and women belonging to wild mushroom–hunting clubs in the Netherlands on their dead reckoning skills by asking them to point back to their car after walking several miles in the woods. Despite their practice at foraging in the outdoors, their accuracy was terrible in comparison to studies conducted among indigenous communities in Australia or Mexico. "From a dead-reckoning point of view," linguist Stephen Levinson has written, "these estimates show that these participants have constructed no clear representation of their current location in the mental map of their immediate environment or integrated that local map into the larger world they know." Rather than navigate their way back, the main strategy of the Dutch mushroom pickers was to pilot in one direction and retrace their steps.

Sea Hero Quest wasn't really created to further science's understanding of how navigational strategies vary across nations or cultures. It was made in order to amass data that will help create a diagnostic tool for Alzheimer's. Spatial ability and memory function are so closely correlated in the human brain that by creating a global benchmark for spatial navigation—what normal is—Spiers and his colleagues hope to be able to make accurate predictions about what a person's spatial navigation performance should be based on their age, gender, and nationality. Doctors typically use language tests to diagnose early-onset dementia or Alzheimer's, but testing a person's spatial cognition performance against these indices could possibly predict even earlier signs of cognitive ailments.

I met Spiers at his office at University College London to talk about the hippocampus and its role in memory. "There are a lot of researchers working on spatial navigation and tasks in rats and mice, thousands of them. They aren't interested in space, they use space to get to memory," he reflected. "I came into the field through memory but then space seemed to become more rich. I've always loved maps and how we actually find our way. It's got this fascinating philosophical element as well. What is space? What is a place?" It struck me that *Sea Hero Quest* cleverly masks a medical test as a video game, but its brilliance is in how it exploits the relationship between space and memory, using one to get to the other.

So *why*, I asked him, did he think navigation is closely tied to episodic memory in the brain? "O'Keefe and Nadel argued that space is something that you can pin things on because it's stable," he posited. "So it's one system, they are totally tied. Space is like a scaffold for adding your memories onto a map."

———

Spiers told me he is often asked whether satellite navigation devices are rotting our brains. His response is that it's important to appreciate the different ways we can utilize these technologies. Using Google Maps on one's phone to figure out a route to get somewhere is not dissimilar from using a print map; following turn-by-turn directions to get somewhere is an entirely different matter. In the spring of 2017, *Nature Communications* published the result of a study Spiers coauthored that tested twenty-four people using GPS to navigate London's Soho neighborhood. It clearly showed that using a GPS navigation system to get to one's destination essentially switches off distinct parts of the brain, including the hippocampus. "Our results fit with models in which the hippocampus simulates journeys on future possible paths while the prefrontal cortex helps us to plan which ones will get us to our destination," Spiers told a reporter. "When we have technology telling us which way to go, however, these parts of the brain simply don't respond to the street network. In that sense our brain has switched off its interest in the streets around us."

Véronique Bohbot does not use GPS. While she was careful to point out that no one has yet designed a study to test whether GPS use causes hippocampal atrophy, there is plenty of evidence that following turn-by-turn directions means we are simply not using a spatial strategy to wayfind. In fact, using a GPS is very much like using the response strategy that exercises the caudate nucleus at a cost to the hippocampus. And because of the remarkable plasticity of the brain, not activating and exercising the hippocampus leads to decreased gray matter. Scientists do know that turn-by-turn directions activate the caudate nucleus, and its response strategy bypasses the creation of cognitive maps. "With GPS you might have even *less* of a reason to pull out that map than we already do," Bohbot said. Lynn Nadel, her doctorate advisor who first connected hippocampal development to infant amnesia, has scoured the data in Bohbot's studies and agrees that the risk of letting our hippocampus atrophy is considerable. "There is a use-it-or-lose-it thing about the brain," said Nadel. "We know on the flip side about taxi driver studies, you can increase the capacity of the system by using it a lot. My gut instinct is yes, if people stop using their brains and totally devote themselves to their handheld devices to find their way around the world, that could have a negative effect on getting around and spillover effects on other things like memory."

Market research shows that the number of turn-by-turn navigation application users reached four hundred million in 2017, a fourfold increase since 2011. Even so, there have been just a handful of studies looking at the impact of using GPS. One of the first was in 2005 when researchers at the University of Nottingham tested twelve drivers who had used either GPS or a traditional paper map, and then measured their landmark, route, and survey knowledge afterward by requiring them to draw detailed maps of their route. The drivers who used GPS remembered fewer scenes, were less accurate, and drew simpler maps with minimal landmarks. The crucial difference between the two methods, the researchers argued, was decision-making—GPS users were not engaged in making decisions.

A couple of years later, researchers at Carleton University conducted a study of 103 individuals and found that using GPS had a number of nefarious consequences for drivers in regard to attention and engagement. Using it replaced direct perception and eliminated the need to gather, integrate, comprehend, and process information from the environment. It also rid drivers of the need for wayfinding, decision-making, and problem-solving. In 2008, Cornell University researchers argued that using GPS reduced a driver's process of interpreting the spaces around them—the process of turning space into place, in other words—while immersing them in a virtual environment, the perspective of the GPS screen. The driver quite literally relies on the virtual representation of the road rather than their unmediated perception of the physical road. Even using GPS while walking seems to change how we move through space: Toru Ishikawa and a team of researchers reported in 2008 that people using GPS while walking did so more slowly, made greater direction errors, and found wayfinding tasks more difficult than those who used paper maps or relied on direct experience.

One academic currently designing a study of how GPS affects the way humans engage with the task of wayfinding is Harry Heft, the professor of environmental psychology at Denison University who studied with James Gibson. "The GPS diffuses that whole way of engaging the world," he told me, "because you don't even really have to *look* at the world very much." GPS has in a way exacerbated changes that were already underway once highways became a predominant medium for travel, he continued.

"The highway system is so disconnected from the terrain and the topology. I think what GPS does is lead you even further away from that."

The vigor of memory is likely one of the first victims of failing to exercise the hippocampus, but it's not the last. We use the neural circuit not only to reconstruct the where and when of the past but also to build images of the future; it is the locus of our imagination. For instance, when asked what he was going to do the next day, the patient H.M. could only manage to say, "Whatever is beneficial." In the 1980s, the psychologist and neuroscientist Endel Tulving also found that his amnesiac patients experienced difficulty imagining the future. Later, in 2007, a series of neuroimaging studies confirmed that the ability to both remember and imagine uses a common brain network that includes the hippocampus. Eleanor Maguire of University College London has proposed that perhaps the hippocampus is not solely responsible for episodic memory, future thinking, and spatial navigation but is necessary for constructing scenes that are crucial to these exercises. Scene-construction theory, as she points out, offers a unified account of why the absence of the hippocampus in individuals destroys so many seemingly disparate functions.

We engage in fascinating gymnastics when we simulate the future—recombining information from our semantic and episodic memory and shaping it into new mental representations of hypothetical events. Our brains are like prediction machines, generating episodes that might occur in the near or distant future and using them to plan, problem-solve, and achieve goals. In this way the human imagination is like a beacon that orients us, helping us to make decisions about where we want to go and how we might get there, as well as self-regulating our behaviors and emotions in the service of a destination or a destiny. Indeed the ability to imagine is a pillar of our autonoetic consciousness, the emergence of which likely set us on our current evolutionary path by extending our identities beyond the present moment into the past and future.

In 2011, researchers Benjamin Baird and Jonathan Schooler showed that we often engage in autobiographical planning and thinking about the future during mind-wandering episodes, those periods when we allow ourselves to fluidly travel between the past and the future. They proposed that mind-wandering enables cognitive operations that are "likely to be useful to the individual as they navigate through their daily life." Tulving described autonoetic consciousness as mental time travel. In the book *Predictions in the Brain*, edited by Moshe Bar, a team of psychologists from Australia and New Zealand point out that our capacity for mental time travel and grammatical language likely coevolved to allow us to share episodic information with one another. This probably occurred during the Pleistocene, a period in which they speculate climate changes necessitated greater social cohesion and future planning, and the prolongation of development from infancy to adulthood—that is, when childhood originated.

In light of the hippocampus's role in future imagining, what are the consequences of decreasing its activity? Could it be that the more we willingly outsource our brain's attention and abilities to a GPS device, the less detailed and more hazy our imagination of the future becomes? Could our societal visions of a collective good, and the steps required to make it so, dissolve into blank spaces too, disorienting us in a far deeper sense than merely losing our way? Navigation, it turns out, may not be something that we excise from our cognition without existential impacts, influencing our ideas about who we are and our destiny.

There are smart people who argue that technologies that offload the cognitive burden borne by individuals are good. In his book *The Future of the Mind*, the physicist Michio Kaku describes a future in which we'll be able to implant memories in our brain, short-circuiting the time it takes to learn new skills and acquire knowledge. For those concerned that such implants would lead to significantly diminished cognitive abilities, that without the necessity of developing important neural architecture for learning and retaining memories and information we might be less intelligent humans, Kaku reassures us that eventually better designed, artificial brains will fix the problem. Kaku even discusses the possibility of implanting artificial hippocampi. Indeed, Theodore Berger, a biomedical engineer at the University of Southern California, has developed a hippocampal implant—a silicon chip that electrically stimulates neurons, and tested it on rats and rhesus monkeys with the aim of improving long-term memory in individuals with Alzheimer's or brain damage. A company called Kernel is reportedly already using it in human trials. In Singapore, scientists have created artificial grid and place cells in software that they claim a robot used to explore an office space. Google has created an artificial intelligence program that can use memory and reasoning to navigate London's underground tube system. But reading about these technological experiments, I found myself revolting against the implicit vision of a technoscientific utopia in which humans can download memories into artificial brains like robots, while robots develop neural networks that think like us. What value will individual experience, practice, and skill have in that future? What pleasure would there be in learning, in childhood, in open-ended exploration, of accidental discovery, of autonomously finding our way? What good could possibly come from outsourcing cognitive processes that were an essential aspect of making humans human? But maybe that future is already upon us.

LOST TESLA

To enter the Harvard Law School Library you walk through a colonnaded facade emblazoned with a phrase in Latin, "Not under man but under God and Law." It was there that I heard a lawyer present a vision for the future of human transportation in which people will travel across the surface of the earth at seven hundred miles per hour in magnetically levitated pods that glide along low-pressure tubes under the power of electric motors, ricocheting commuters between cities in minutes, bending the laws of nature to the will of man. The lawyer worked for the company Hyperloop One, and the idea had been first described by Elon Musk, founder of the luxury electric car company Tesla, as a cross between a Concorde, a railgun, and an air hockey table. Imagine something like the plastic tube used to shoot checks from your car to the bank teller. On the Hyperloop, a trip from Melbourne to Sydney, normally eleven hours of driving in a car, would take fifty-five minutes, and the physical experience of traveling would be of initial thrust followed by zero sense of motion at full speed. The company's motto is "Be Anywhere, Move Everything, Connect Everyone," and they already have a test track in Nevada. The United Arab Emirates has a deal with the company to build a train from Dubai to Abu Dhabi that would travel at five hundred miles per hour and carry passengers between the two cities in just twelve minutes. (Virgin has now invested in the company and it became Virgin Hyperloop One.)

Like the invention of the wheel, car, train, and airplane before it, technologies like the Hyperloop and future commercial space ventures will realign whole economies and patterns of movement. They may also realign the human mind, much like the invention of the map, the view from a plane, or a photograph of the earth from space did. However, in an airplane, a train, and even a driverless car, a passenger still has a perspective—a window through which to perceive: Hyperloop promises travel inside a toilet paper roll devoid of any visual reference to one's surroundings or reminder of our fleshly selves moving through space, time, life. What might we lose when we relinquish any view at all? In his book *Skyfaring*, the commercial airline pilot Mark Vanhoenacker describes the window seat of a plane this way: "Whatever our idea of the sacred, our simplest questions—how the one relates to the many, how time equates to distance, how the present rests on the past as simply as our lights lie on each night's darkened sphere—are rarely framed as clearly as they are by the oval window of an airplane." The future might be one in which we have fully turned inward even as we roam vast virtual worlds of our own creation.

Looking back at technological ambition in the twentieth century, so many of them have focused on making the world easier to access, hurling people farther and faster while requiring them to make the least amount of effort possible. Early on, some were concerned about the effect on our souls. When Anne Morrow Lindbergh boarded a commercial flight from America to Europe in 1948, she wrote about the experience for *Harper's Magazine* and said that airplane travel produced an illusion of terrible power and freedom in passengers. They could observe the earth beneath them even as they were insulated, detached, "comfortable, well-fed, aloof, and superior." Airplanes shrank the globe while inflating our sense of scale and power over it. A few years before her, Wendell Willkie wrote that "[t]here are no distant places any longer; the world is small and the world is one." Around the same time, Henry Luce offered his vision of American globalism, which could be the motto of the modern, entitled, globetrotting class today: we have "the right to go with our ships and our ocean-going airplanes where we wish, when we wish, and as we wish."

Our love affair with speed is long-standing. The futurist Filippo Marinetti wrote in 1909 that "the splendor of the world has been enriched by a new beauty—the beauty of speed." One writer in the magazine *Aeronautical World* breathlessly reported in 1902 that "[a]s the speed of aerial transit may reach several miles a minute man will practically be able to annihilate space and circumnavigate and explore the whole surface of this globe with independence, ease, depth and economy, or travel from pole to pole, or where ever his fancy may dictate, unhampered by restrictions of any kind." When the Concorde took its first flight in January 1976, it was the apotheosis of human travel, allowing us to move at supersonic speed and sip champagne at the same time. Charles Lindbergh, Anne Morrow's husband, died two years before the Concorde flew, but he had already predicted that the perfection of machinery would insulate man from contact with the elements. "The 'stratosphere' planes of the future will cross the ocean without any sense of the water below," he wrote in the foreword to his wife's book *Listen! The Wind*. "Only the vibration from the engines will impress the senses of the traveller with his movement through the air. Wind and heat and Moonlight take-offs will be of no concern to the transatlantic passenger."

Now the days in which flying an airplane meant reckoning with metal, combustion, temperature, horizons, and physics seem as far away as the creak of a horse and cart or the gush of a hand-pumped well. We're more eager than ever for technologies that offer faster, time-saving, insulating travel, not so much as machines as swaddle blankets protecting us from the wild uncontrollability of the world outside our heads.

Time—saving it, managing it, maximizing it, escaping it,

denying it—strikes me as one of the prime concerns and anxieties of modern life, and hence a metric by which we judge the quality of our travel. How fast is it, how easy to distract ourselves from the means and act of it, the burning of fuel, the people and places we fly over? "We're not selling transportation, we're selling time," promises Hyperloop.

Perhaps the World Wide Web already represents that achievement, allowing us to get rid of the need for travel at all. Alexander Graham Bell, inventor of the telephone, predicted that a day would come when we would be able to see the person we were talking to over the phone. Sure enough, we now experience that miraculous instantaneousness via our phones and computers. Beamed over fiber-optic cables and satellite waves, we can contract the entire world. I find it telling, though, that while the web rids us of the need to physically travel from one place to another, we cling to navigation as a metaphor for the act of being online. Confronted with virtual space, we still grasp for the structure of real space, describing the web as a place we locomote through even though it doesn't exist anywhere. We *search* for content and go *forward* and *back* between *sites* that we *visit*. The icon for the web browser Safari is a compass.

Hypermobility has enabled our individual range and consciousness to stretch across the entire surface of the world. Yet our mastery is uncomfortably flimsy. It sputters and dissipates the moment the gas runs out or the battery fades. It seems to me now that our range of motion without the crutches of technology has actually shrunk, and our intimacy with the places we go may have dimmed.

In places that have long-established traditions of navigation by environmental cues, GPS can represent yet another onslaught against cultural identity. I watched the filmmaker and *Hōkūleʻa* crew member Nāʻālehu Anthony hold up his smartphone in front of an audience and tell them, "The compass and the sextant and the

GPS. This device can co-opt three thousand years of knowledge by pressing a button and looking for the pathway." When the anthropologist Claudio Aporta began studying Inuit wayfinding in the Canadian Arctic, he wondered whether GPS was just another technology that communities in the Arctic would adapt to and thrive with, like snowmobiles or shotguns, or would it erode something intrinsic and crucial about Inuit culture itself? When he first went to Igloolik in the 1990s, some forty hunters already owned GPS units. The device's greatest benefits were during walrus hunts: hunters could save fuel returning to shore from their hunting sites by plotting a direct course even when the shore was out of sight. But those who had grown up on the land still didn't use GPS much, and knowledgeable full-time or part-time hunters merely used it to supplement traditional wayfinding. It was younger hunters who tended to rely the most on GPS as a primary tool. The combination of a lack of wayfinding experience, the speed of snowmobiles, and the ease of GPS could quickly amplify the dangers of navigating in the Arctic. GPS changed the routes that people take, sometimes away from paths whose safety had been proven over generations; some hunters can tell just from observing tracks in the snow who was using GPS to find their way because they were straight as an arrow—a computer-plotted track. Jason Carpenter, teacher at Nunavut Arctic College, told me that "[i]t's easy for anybody to jump on a skidoo and get out a hundred miles almost without thinking. So our ability to get ourselves in a bad situation is greater."

Many of Igloolik's residents who knew the most about traditional wayfinding were in their seventies or eighties, members of the last generation who had been born on the land; had been schooled in wind direction, snow, sun, stars, tides, currents, and landmarks; and had memorized hundreds of place-names. After GPS arrived, hunters could minimally rely on the environmental cues, and it lightened the cognitive load of memory itself. "The GPS receiver's answer to a spatial question (e.g., where to go) is provided by a mechanism that is physically detached from it (a network of satellites) and required no involvement of the traveler with the environment," Aporta and his coauthor, Eric Higgs, wrote in their paper "Satellite Culture: Global Positioning Systems, Inuit Wayfinding, and the Need for a New Account of Technology." "Although the act of physical travel will always involve some connection with the surroundings, this connection is . . . shallow." In Igloolik, Alianakuluk, an elder, told Aporta about a rescue operation in which the searchers wanted to use GPS to follow a course. He knew, however, that it would lead them straight into a dangerous landscape and the floe edge. "I told him that I better lead the way and I will lead with Inuk knowledge, otherwise we would get to the rough pressure-ridges field. So I led after that, using snowbanks created by the prevailing *uangnaq* wind . . . as my wayfinders," said Alianakuluk. "We did reach our predetermined destination using my knowledge as an Inuk. Had we just followed the GPS we would have gone through rugged pressure ridges, then even possibly to the floe edge. This would have caused more problems than help anyone. That I know for a fact."

We are all neophytes when it comes to GPS, computers, the World Wide Web, and jet travel. These are only barely newer to Western societies than they are to indigenous ones. "GPS is basically having an effect on how we relate to space and geography in general, due to the fact that spatial decisions we used to make on our own are now made with a device," Aporta explained to me. He cited the work of the philosopher Albert Borgmann, a professor at the University of Montana. Since the 1980s, Borgmann's work has focused on a theory he calls "the device paradigm" that seeks to explain the ramifications of technology at the personal, social, and political levels of modern existence.

Nearly every aspect of human life, says Borgmann, has been affected by the replacement of things with devices. Craftwork by automation, candles by lighting systems, fire by central heating. Devices can do many things, including releasing us from darkness, cold, and hardship, but they also separate people from the physical environment by subordinating nature. So while devices liberate people from toil, freeing our time and energy, they also separate the means from the end. We are disconnected from the environment and the skills required for daily survival. Consider a thermostat: it allows us to control the temperature of our homes with a finger, yet by using it we are no longer responsible for physically gathering the resources needed to heat our own homes—the thermostat conceals the means of heat. According to Borgmann's argument, the divorce that devices create cumulatively erodes social and ecological meaning.

GPS is the perfect Borgmannian device. Even though it had not been sold on the mass market yet, the philosopher might have been describing it when he wrote in 1984 that "the machinery makes no demands on our skill, strength, or attention, and it is less demanding the less it makes its presence felt." Of course, navigational devices like maps, compasses, and sextants also fit Borgmann's device paradigm, because they outsource to some degree the formidable experience, observation, and memory needed to undertake skilled navigation. But even these inventions required a level of environmental awareness and orientation, as well as an understanding of topography or celestial phenomena. It wasn't until the twentieth century that navigational technology released us from needing to pay *any* attention at all. "The combination of newer navigational instruments (e.g., radar, automatic beacons, computational support) produces an increase in efficiency and a corresponding loss of skill," write Aporta and Higgs.

None of us is exempt from the ramifications of the device paradigm. We all seem to find it extraordinarily difficult to step outside the onslaught, to create the distance and perspective between us and our devices that might allow us to question what cultural or cognitive price is being paid in return for convenience.

———

The French author and pilot Antoine de Saint-Exupéry hated technophobes who attacked machines as the source of mankind's ills. He thought that machines would become a part of humanity rather than a foe because they would connect us to one another. "Transport of the mails, transport of the human voice, transport of flickering picture—in this century as in others, our highest accomplishments still have the single aim of bringing men together." And rather than act as mediating objects that divorced man from nature, machines to Saint-Exupéry were devices that could bring us *closer* to nature. "It is not with metal that the pilot is in contact," he wrote in *Wind, Sand and Stars*. "Contrary to the vulgar illusion, it is thanks to metal, and by virtue of it that the pilot rediscovers nature. The machine does not isolate man from the great problems of nature but plunges him more deeply into them."

What, I wondered, would Saint-Exupéry, who died in 1944, make of autonomous vehicles? The race to fill the world's roads with driverless cars is well underway. Some ten million are predicted to be in use by 2020. Google, Mercedes, BMW, Nissan, and Tesla are just some of the global companies already testing prototypes. These autonomous vehicles are about as different from the steel machines of the early twentieth century flown by Saint-Exupéry as a violin is from an iPod. Equipped with lidar, radar, sonar, infrared, and ultrasound, the cars "sense" the environment around them. Some also merge the data they are collecting with three-dimensional maps of the environment stored on the digital cloud. Because some autonomous cars require these premade three-dimensional maps in order to navigate safely, cartography is poised to undergo yet another revolution. The maps used by driverless cars won't just represent every yard but every *inch* of the environment in order to tell the difference between a tree and a child. Yet another effort to map the earth in this level of detail may seem innocuous. We already map galaxies, the brain, the bottom of the ocean, and the surface of Mars. We can already "drive" down a street using Google Street View and access Google Earth to gain an allocentric perspective of the surface of the world. The mapping technology required for driverless cars to work may not appear to present any more radical a leap. But I see potential for a nefarious influence on our lives: the more we rely on driverless cars and their need for three-dimensional maps, the narrower our choices of where we travel and explore could become. As we grow to depend upon, or maybe even just prefer, autonomous vehicles, they will choose the routes we take, and they in turn will choose those routes that are already mapped. Where we go will increasingly be confined by the technology we use.

The benefit of self-driving automobiles is, we are told, that they are more precise, reliable, and therefore safer than human-operated cars. They will allow us to travel at faster speeds, potentially even bumper to bumper, and rid metropolises of pollution-causing traffic and congestion. They will eradicate the need for parking in cities by delivering passengers and returning later to pick them up again, thereby transforming public and private land use. Autonomous vehicles could reduce CO_2 emissions if they are used for ride-sharing. But is the future of autonomous vehicles really so utopian? Driverless cars could also end up exacerbating the very problems we already have with modern transportation systems. People will be willing to commute longer distances, *ballooning* air pollution and CO_2 emissions. Autonomous vehicles might not be used as ride-sharing tools but as mobile spas in which people are pampered as they sit back and let the car do the work of getting them to work and home. "If I can go a hundred miles an hour bumper to bumper, I can live in the Berkshires and be just as late to the office now as when I live eighteen miles outside Boston," said Joe Coughlin, a transportation expert at MIT. "When we're talking about autonomous vehicles, we're talking about how do we want to live together as a society?"

The allure of autonomous vehicles is that so much of our travel along monotonous highways slashed across the landscape, often bordered by sound-insulating walls so as to limit one's view, already feels like a waste of time. If our brains are on autopilot while driving, responding to GPS-simulated instructions, wouldn't it be better to spend that time in some other way? Why shouldn't driving become the next flying, where we sit back and relax as we are taken to our destination? But I see the driverless car as representing yet another severance between our movement through space and time and the pleasure of effort and autonomy. GPS relieves us of the need to form cognitive maps, and driverless cars relieve us of the need to look away from our screens and take note of directly experienced phenomena in the environment. In seeking maximum speed and ultimate efficiency, the autonomous vehicle cocoons us from the physics of movement at high speeds and the burning of fuel. Rather than plunge us into reality, an autonomous vehicle erases it.

Maybe I'm wrong. Perhaps the autonomous vehicle is the arrival of the perfect machine, what Saint-Exupéry described as the one that "dissembles its own existence instead of forcing itself upon our notice." Maybe it will allow nature to resume its pride of place, more deeply connecting us to our surroundings. But I worry that they will further embed us in our individual diversions. Consider the tragic story of Joshua Brown, the first person to be killed in an autonomous vehicle. Brown, a forty-year-old Tesla enthusiast, was speeding at seventy-four miles an hour down a Florida highway using the car's autopilot feature when its sensors failed to detect the white broadside of a truck against a bright blue sky. The car smashed into the side of the truck, which sheared off the car's top and killed Brown. What was he doing at the moment of impact? Watching a Harry Potter movie. We know this because the truck driver could still hear it playing on a portable DVD player after the car sped three hundred more feet and hit a telephone pole so hard it snapped in half.

———

Hundreds of thousands of years ago something brilliant occurred in the human brain. A circuit was completed, a spark of inference caught fire, and our consciousness became unique in the whole of history. We went seeking and found our way back. We invented time to measure the distances we traveled. We recognized in the ground a record of the past that we could reconstruct, and we traveled forward in time to imagine a not-yet-real moment in the future. We became creatures capable of abstract thought. We invented narratives with a beginning, middle, and end. The first epics may have sprung from the mysterious firings of our hippocampal cells that proliferated as we strove to travel longer and longer distances. We discovered we could describe out loud the things we saw and the places we wanted to go, we could tell each other stories, and we began to traverse the world in the likes of migrant birds and animals, changing it as we did.

Navigation may have served as an evolutionary precursor to storytelling, but humans quickly started to use storytelling as a wayfinding tool, one that helped us have the greatest geographic distribution of any species. The human mind seems built to encode topographical information in the form of stories. We created repositories of shared memories in some places and developed deep, emotional attachments to them. We called those places home. We accumulated knowledge of nature through observation, intimacy with the sun, moon, stars, wind, and landmarks and created complex traditions of practice to know where we were, to seek out places, make new homes, or return to ancient ones. The same brains that evolved to encode space and time shot new navigational guides into space in the form of satellites. Stories proliferated, a noosphere of human experience and memory ensconcing the world.

From a Darwinian point of view, wayfinding is a condition of survival: we do it to avoid predators, to find food and shelter. But understanding *why* we move today isn't only about the methods and conditions. There are deeper levels of the mind and soul to consider, the internal things that repel, pull, call, and force people to go places. We can't reduce human wayfinding to survival because it fails to meaningfully explain the full spectrum of our experience, our fears, dreams, and hopes that drive us—the traits that make us human. It's these traits that seem most imperiled in a future of handheld GPS devices and self-driving vehicles, a future in which we willingly give up autonomy in order to avoid getting lost.

In the fall of 2016, I started following a Twitter account called @lostTesla. It was created by Kate Compton, an American computer programmer who told me she was on a drive from Boston to New Hampshire when she started thinking about a tweet from Elon Musk. He wrote that at the tap of a button on your phone, you can summon your Tesla, which "will eventually find you even if you are on the other side of the country." Something about the "eventually find you" captured Compton's imagination, and she wondered about a car's internal experience. What does it perceive? Does a car possess qualia, meaning an individual subjective, conscious experience? At a hotel that night, Compton wrote a bot—software that autonomously tweets based on a programmer's code—in an hour or two. She imagined the bot to be a Tesla car that had become lost and was wandering, driverless, through an unknown American pastoral setting in search of its human owner. Since then, every day twice a day, @lostTesla has been tweeting its thoughts:

what is a sparrow

I leave a quiet town / now a silo, a silo, a silo / the sun is gone now. Everything is gold / my sensors detect light.

i notice my reflection in the windows of a department store. i'm covered in flower petals

My hood is wet / i see a shiny truck. Many chickens. Can I be a chicken with you? SetMode:FEELING_PRESENT.

I stop to recharge. will i dream? i dream of rabbits. toggleFlag:DREAMING

Over the months that I followed @lostTesla on its travels, the car seemed to become less of a machine and more of a smelling, hearing, seeing, embodied entity. Its movements in space provoked marvelous observations and a seeming surprise and pleasure at the depth of its own feelings toward the world. Maybe this is a logical evolution. The experience of being lost is uniquely human. Animals, equipped with biological instruments that seem to give them absolute certainty about their geographic position, rarely seem to. It is humans who have had to evolve the intellectual and emotional capacity to solve the problem of becoming disoriented and who generated the cultural practices of wayfinding. Maybe our emotional connections to each other and places serve navigational purposes, in the sense of orienting us over very large spatial and time scales—the journey of a lifetime. But in the future, the problem of being lost could become an increasingly rare experience, one that our descendants may deign an unfortunate weakness of human cognition and a quirky artifact of history, even as Lost Tesla, still roaming the landscape, discovers its own joyful sentience.

EPILOGUE
OUR GENIUS IS TOPOPHILIA

The first thing John Stilgoe, a landscape historian, told me was, "I feel sorry for your generation. It doesn't get lost much." I sat across from Stilgoe in his office at the top of Sever Hall in the northeast corner of Harvard Yard. Dressed in a wool suit and bow tie, he has been holding court here for decades as a professor in the Department of Visual and Environmental Studies, and he is known as a polymath whose intellectual range spans history, transportation, fashion, literature, ecology, and the pleasures of bicycle riding.

I was there because one of Stilgoe's central concerns is what he calls visual illiteracy. Busy, rushed Americans, he thinks, no longer take the time to explore and discover their surroundings and have lost their capacity to even see them directly. Stilgoe believes that in an age dominated by programmed, mediated material and the internet, the ability to *look* at things, to practice the art of observation that was required of us as a species for eons till now is an essential aspect of our intellect and has to be relearned. To this end, Stilgoe teaches a course called "Scrutinizing the American Environment: The Art, Craft, and Serendipity of Acute Observation," which exposes his students to thousands of images taken from American suburbs, farms, industrial zones, and recreation and abandoned areas in an effort to permanently alter the way students see the world.

In his book *What Is Landscape?* Stilgoe writes that "[a]nalyzing landscape empowers. Noticing—noticing without keeping any sort of journal, visual or otherwise—reveals and entices. Piecing together what one sees when one wanders or walks quickly on everyday errands requires only will and practice. . . . Exploring landscape, however casually, is a therapy and magic of its own. But it depends on curiosity and scrutiny." Stilgoe has dedicated entire books to answering the question of what constitutes a landscape in the first place, settling on a definition of the surface of the earth shaped by humans. He suggests that the presence of navigational aids, whether they are reefs in the sea or topographical features on land, can transform wilderness into shaped land, that is, into landscape.

On this day in particular, Stilgoe's concerns are with my generation, and the ones to come, for whom GPS is a natural wayfinding tool. To use it, you already have to know where you want to go, making it the enemy of wandering. Exploration—preferably on foot but also by bicycle, canoe, horse, or ski—is, according to Stilgoe, the best way to stretch the mind, because it facilitates discovery. And for someone who prizes exploration as a primary method of human insight and worships the kind of ambulatory exercise that allows the mind to wander, there could be no more dystopian future than one in which humans relinquish their unique capacities in return for mere efficiency. To him, getting lost is an opportunity for discovery, one that demands that all the senses come alive, and creates a maximum alertness in which observation and possibility are heightened. In *What Is Landscape?* Stilgoe writes that "[b]eing lost, even being deliberately free of electronic location devices, sharpens one's senses and often eventually reassures. Making one's way often reveals paths distinct and well used or hard to discern, abandoned (perhaps for good reason), but all nonetheless instructive."

"If I'm lost and I don't have anyone to ask, I love that feeling," he said, though he drew a distinction between being desperately lost in dangerous circumstances and getting lost in a generally unknown place. In the latter case, to go off track is really about challenging the borders of one's familiarity, pressing beyond the known spaces of our understanding and experiences and into the new. "There are categories of being lost. Panic means you have the voice of Pan in your ear," he said.

Stilgoe grew up and still lives in the seaside town of Norwell, Massachusetts, where his father was a boatbuilder and his mother a homemaker. He had free rein of the woods, marshes, coast, and sea. Nowadays, Stilgoe lamented, children don't venture into the marshes around his hometown, even after public assurances by the local police that these spaces are perfectly safe. Girls especially, he feels, have been cheated by the lurking terrors within the American psyche, the nearly hysterical response to perceived risk. To see the true extent of change over the generations you have to go back a bit. Over the last century, but particularly the last few decades, Stilgoe believes society has circumscribed freedom of movement in America, especially for children. Stilgoe has documented how in the 1890s children and teenagers engaged in national recreations like canoeing, bicycling, and even amateur ballooning or flying with large, tethered kites. "Men and boys built gliders, lay beneath them, and coasted down snow-covered hills until the airfoils lifted them from careening sleds," he wrote. How many teenagers today would be content to walk alone outdoors? And if they did, would they have the vocabulary to describe what they saw? "Exploring, being lost for a while, looking around without distraction, or just going for a walk eventually raises questions of words, if only in the telling and retelling of short-term adventure," Stilgoe writes.

Part of the problem is parents' incessant management of their children's time. "I think they've missed a kind of self-guided, nonorganized activity, nonsports activity growing up. Wandering around, getting into things. And the assumption seems to be nowadays is if a child isn't in an organized activity, the child is a criminal," Stilgoe said. "But as far as I can understand, most of my colleagues I work with seem to have found their careers by being slightly disorganized. Lucking into something, you know?"

Changes in the way kids get around are also part of the problem. Bicycles used to empower children to engage in short-range exploration of local worlds until, he said, ten-speed bicycles, whose chains are prone to being caught in brush, steered them out of the old fields and woods and onto roads. Today's kids might not even be allowed to ride bicycles on streets without supervision. And there are other, more nefarious fears at work preventing children from roaming by themselves. American landscapes today are infused with perceptions of menace and a sense of dread, creating boundaries around possibility and limiting the places where children are free to go.

Studies of contemporary children's "home range," the distance from home that children are allowed to go outdoors, show a dramatic decrease in the "right to roam," not just for kids in the United States but also those in Australia, Denmark, Norway, and Japan. One 2015 study in *Children's Geographies* that was focused on Sheffield in north England showed how over three generations the children in a single family had undergone a severe hemming of movement. The grandfather recalled that he could travel several kilometers without asking permission to fish, ride bikes, visit friends; the only limitations were the weather and hunger. "If we were on our bike [parents] never [knew] where we were." Meanwhile the second-generation parent was only allowed to travel half a mile from home without permission. And the third-generation child wasn't allowed to go anywhere without permission and was only allowed to go one place with permission: a friend's house three doors from home. A newspaper story published in the *Daily Mail* in 2007 told a similar story. Over four generations in Sheffield, children had gone from being able to travel as far as six miles from home without supervision to no distance at all; one child was driven in the car everywhere, including to the playground nearby that his mother had once been allowed to walk to on her own. As the study's authors point out, the consequences of these changes are multifaceted, impacting physical and social skills. And, they write: "Autonomy is a key to the acquisition of spatial skills, therefore the development of these skills can be hindered if children cannot move independently in the outdoor environment."

For Stilgoe, the ubiquity of smartphones can never be a good thing. His career has been dedicated to creating thirst in his students for curiosity, exploration, and wonder—believing them to be in fact the keystones of true intellect. Smartphones don't open the user up to their surroundings, they funnel attention into themselves and a universe where everything is known, mapped, and accessible. "I'm grateful I didn't grow up with a smartphone," he told me. "My students don't even know why I'm grateful."

He paused and stared at the ceiling. "Oh, I'm a bundle of joy."

***

The French sociologist and philosopher Pierre Bourdieu described the world as a book that children learn to read through the movements and displacements their bodies make in space. Through their motion, children create the world around them as much as they are shaped by it. Of course, the first place a child experiences is the womb. The womb is not a void but a place of many sensations: a fetus hears sound, perceives light, smells, and tastes. Swimming in amniotic fluid informs the development of a nervous system. When he first emerges, a newborn baby is *worldless*, meaning the borders between himself and his surroundings are nonexistent. In the first weeks and months, he seeks out those boundaries, where skin and object meet, using mouth and touch to begin to establish spatial experience and knowledge of his new reality. At birth, Jean Piaget wrote, "there is no concept of space except the perception of light and accommodation inherent in that perception. All the rest—perception of shapes, of sizes, distances, positions, etc.—is elaborated little by little at the same time as the objects themselves. Space, therefore, is not at all perceived as a container but rather as that which it contains, that is, objects themselves." By exploring and moving, alighting their hippocampal cells, infants create spatial representations in the brain and build the architecture for episodic memory, the crucial component of autobiography and sense of self through time.

As infant amnesia erodes and the fleeting capacity to retain memories strengthens, other remarkable characteristics start to emerge in children. They develop personalities and the ability to create private worlds and stories, which become powerful generators of intelligence and knowing. In 1959, the American psychologist Edith Cobb called these capacities the "genius of childhood," thinking that they allowed children to develop intense bonds to places. A close friend of anthropologist Margaret Mead, Cobb was fascinated by why childhood is so crucial to human evolution and culture. She defined childhood as a period from approximately five or six to eleven or twelve years old, and argued that the gift of this prolonged childhood in contrast to other species was the profound plasticity it gave children to respond to their environment. "This plasticity of response and the child's primary aesthetic adaptation to environment may be extended through memory into a lifelong renewal of the early power to learn and to evolve," wrote Cobb.

For her research, Cobb consulted some three hundred volumes of autobiographies, searching some from as far back as the sixteenth century for accounts of childhood. The results led her to believe that children are geniuses in a rather specific sense; she used a definition of genius in its early iteration, as "the spirit of place, the *genius loci*, which we can now interpret to refer to a living ecological relationship between . . . a person and a place." It's particularly during the middle period of childhood that Cobb thought children experience the natural world in highly evocative ways, when they begin exploring a new awareness of themselves as having a separate and unique identity in relationship to the outside world. They gain an exceptional perception of time and space, and profound moments of transcendence from its continuum. Cobb felt that places—concentrations of meaning, intention, and experience—spur children's sense of self.

Cobb was not the only one to recognize that children have a unique capacity for strong bonds to places. The psychologist James Gibson wrote that a "very important kind of learning for animals and children is place-learning—learning the affordances of places and learning to distinguish among them—and way-finding, which culminates in the state of being oriented to the whole habitat and knowing where one is in the environment." The French geographer Eric Dardel wrote that "for man, geographical reality is first of all the place he is in, the places of his childhood, the environment which summons him to its presence." Children inhabit and experience places as having existed before the capacity for choice, before they develop their separate identity. They feel primordial, as though they have existed since the beginning of time. "Before any choice, there is this place which we have not chosen, where the very foundation of our earthly existence and human condition establishes itself," wrote Dardel in his book *L'Homme et la terre*. "We can change places, move, but this is still to look for a place, for this we need as a base to set down Being and to realize our possibilities—a *here* from which the world discloses itself, a *there* to which we can go."

So many cultures cherish the metaphor of a road or journey to describe life; where we are born is the starting point of that epic. Often the places we grow up in have outsized influence on us. They influence how we perceive and conceptualize the world, give us metaphors to live by, and shape the purpose that drives us—they are our source of subjectivity as well as a commonality by which we can relate to and identify with others. Maybe it's because of the vividness of their sensory impressions, their genius for establishing deep relationships to their early environments, that children have a strong capacity for the human emotion called *topophilia*. First defined by the Chinese-American geographer Yi-Fu Tuan, *topophilia* is the sense of attachment and love for place. In his book on the subject in 1974, Tuan describes topophilia in universal terms.

> Of course, peoples of the desert (nomads as well as sedentary farmers in oases) love their homeland; without exception humans grow attached to their native places, even if these should seem derelict of quality to outsiders. . . . As a geographer, I have always been curious about how people live in different parts of the world. But unlike many of my peers, the key words for me are not only "survival" and "adaptation," which suggest a rather grim and puritanical attitude to life. People everywhere, I believe, also aspire toward contentment and joy. Environment, for them, is not just a resource base to be used or natural forces to adapt to, but also sources of assurance and pleasure, objects of profound attachment and love. In short, another key word for me, missing in many accounts of livelihood, is Topophilia.

Tuan's definition of topophilia is, I think, germane to wayfinding. Across cultures, navigation is influenced by particular environmental conditions—snow, sand, water, wind—and topographies—mountain, valley, river, ocean, and desert. But in all of them, it is also a means by which individuals develop a sense of attachment and feeling for places. Navigating becomes a way of knowing, familiarity, and fondness. It is how you can fall in love with a mountain or a forest. Wayfinding is how we accumulate treasure maps of exquisite memories.

———

Mau Piailug was a toddler when his grandfather on the island of Satawal first began putting him in tide pools to feel the ocean's pull and push. Solomon Awa was a baby when he began traveling by dogsled with his parents from camp to camp. Bill Yidumduma

Harney was raised in the bush and spent the nights of his child-hood staring at the stars and learning the stories of their move-ment. In each of these examples, the practice of observation, the education of attention, began early in life, a process of attuning perception to the environment, committing story and intergen-erational knowledge to memory. This process is one that Bourdieu might have called *habitus*, the orienting of human behavior through the transmission of practices, "a whole system of predis-positions inculcated by the material circumstances of life and by family upbringing."

Today the conditions of modern life and technology have changed the skills and knowledge required for survival, and those things that are not learned and practiced are eventually lost. "Everything you don't practice is a lost skill," David Rubin, a neuroscientist at Duke University and expert in oral traditions, told me. "People used to build wagon wheels. That's gone. No one can fix automobiles. I have a car and I can't check the oil anymore. Things are changing. If we don't sing ballads, they will be lost. But that doesn't mean we aren't still *capable* of doing it."

While the mastery of navigation in traditional cultures often begins early in life, I discovered that it is never too late to begin learning. And starting the process is extraordinarily simple. It re-quires no journeys to faraway places or money. It can be as simple as going outside to direct your attention to the environment. It might be the difference between looking down while walking and looking up. It can start with practicing acute observation of the places you already live.

I frequently asked people I interviewed about navigation for their advice on how to accumulate those skills or improve mem-ory, or I searched for answers in their work. Again and again, I was surprised by how simple the answers were. "Learn to draw," said Rubin. "We don't know how to represent the world well enough. Actually paying attention to the environment, making empirical observations and organizing them into a system—do that." Tristan Gooley, the British expert in natural navigation, advises that we focus our powers of deduction on the natural world. Harold Gatty recommended going for a walk, preferably alone, and "think[ing] purely of the external world. The man who walks to solve an internal problem, to ease his mind, or to daydream, is going to learn nothing about natural navigation," he wrote. "Eventually small hills, stones, trees, bushes, are mentally recalled with very great ease, and in their proper succession, and become bound in the observer's memory as the links of a chain."

Véronique Bohbot at McGill University has found that it takes just a couple of months of twice-weekly spatial memory exercises that gradually increase in difficulty—proceeding from recalling the positions of objects in rooms to navigating museums, say—for people to increase gray matter in the hippocampus, and she has created a program called VeboLife that is like physiotherapy for this region of the brain. "We're teaching people to look at the environment," she said. She recommends that people interested in cognitive health make changes to their daily routines by incorporating new behaviors into their life, like taking new streets and shortcuts, creating mental maps. John Stilgoe advises his students and readers to start with looking around and "ask always the names of what comes to mind as one walks slightly inland, to look under bridges, to walk in the dark, to ask about color, to always think about what it means to fly as contemporary airline passengers fly, not as teenagers once flew, to find lunch and remember that the food came from a farm, usually a farm in fly-over land, and to think always of home and whatever home means."

James Gibson believed that we could reeducate our attention. We think of ourselves as existing inside our heads, separate from the world, but he thought we could see it directly and even share those unmediated perceptions with others. The philosopher Albert Borgmann encourages individuals to become cognizant of the creep of technological change into the important centers of their lives, to consider what he calls "focal things." Focal things and the practice of them require exertion, patience, engagement, skill, discipline, fidelity, and resolve; they engage the body and

the mind because they have a commanding presence, they demand our attention. Focal things might be a hearth, a meal, carpentry, a craft, or hunting, and the practice of them might be
gathering wood, cooking, building, making, or tracking.

Over the course of writing this book, I made navigation a focal thing, diverting my attention to how I did it, taking note of
my surroundings, filing them away in my memory. I took up the
habit of wearing a small compass on my wrist to use in new or
old places, allowing me to infer directional information from
buildings, waves, wind, or trees that I observed. Eventually the compass became less important to divining this information. I took a
small journal everywhere so I could jot down little, seemingly insignificant things that I noticed, cultivating a kind of practice of
observation, and I tried to notice something at least once a day,
though sometimes it seemed like a whole week could go by before I stopped to look around. "On the way to Joaquín's school is a
towering tree with elephant skin–like bark whose leaves and buds
are just coming in. The leaves are pale yellow-green and the buds
droop like streamers or pompoms. I think it is a boxelder maple,"
I wrote one day. I strove to know better the five-hundred-acre park
across from my home in Brooklyn by pushing into less-familiar
places, and at a scale of intimacy I hadn't considered before; a grove
of sycamore trees or the patches of yarrow I had never stopped to
look at.

Even in my own neighborhood I found that I could navigate
the "border-land of knowledge into the realm of the undiscov-
ered," as the naturalist and educator Anna Botsford Comstock has
described. In her 1911 book *The Handbook of Nature Study*, Comstock said that "the study of nature consists of simple, truthful observations that may, like beads on a string, finally be threaded
upon the understanding and thus held together as a logical and
harmonious whole." Comstock's nine-hundred-page book was an
offering to public school teachers and parents, but especially
children, who she felt were living in an age of nervous tension and
diminished freedom and were at risk of losing the powers of ac-

curate observation, the practical and helpful knowledge that nature provides freely. She wanted children to cultivate through nature study the "perception and regard for what is true and the power to express it."

Modernity has brought intense turmoil and change to how we move and the reasons we move. Whether you see this as a positive or a negative development might have a lot to do with how much autonomy, safety, and freedom you enjoy in deciding when, where, and how you get from A to B. While some have gained access to the globe and the ability to travel across it, sometimes on a whim, others are forced into upheaval against their will. The future seems to promise even more of this roiling disruption for the vulnerable among us. The International Organization for Migration reported that in 2015 there were more international migrants than ever recorded before. Some 244 million people now reside in a country other than the one they were born in, and the phenomenon of forced displacement has risen 45 percent in just a few years, fueled by refugees, asylum seekers, and internally displaced people in Africa, the Middle East, and South Asia. Thirty-eight million people alone have been pushed to leave their homes within the borders of their own country by conflict and violence. Climate change is sure to bring increased flight and migration. Even the most conservative estimates predict tens of millions of climate refugees by the year 2050.

At the very same moment these upheavals occur—or perhaps in response to them—society seems ever more determined to regulate people's movement, to close off certain places through passport controls and physical barriers. Political scientists Ron Hassner and Jason Wittenberg say the number of fortified borders between countries has massively increased since World War II. In the 1950s there were just two, but over subsequent decades border walls have steadily grown: since the fall of the Berlin Wall, the *Economist* reports, forty countries have built walls against more than sixty of their neighbors. Half of the fifty-one boundaries built since World War II were built in the period between 2000 and 2014; often it's wealthy countries trying to prevent people from poorer countries from coming in. Should freedom of movement be an explicitly designated human right? "In the context of massive inequality, the current border regime is even more unjustified, akin to the arbitrary and anti-human character of a global caste system," writes political scientist Guy Aitchison.

The consequences of mass upheaval are the fracturing of communities and the severance of roots that connect us to places and each other. In *The Need for Roots*, the French philosopher Simone Weil claimed that "to be rooted is perhaps the most important and least recognized need of the human soul." A condition of life is that each human being has multiple roots in order to "draw wellnigh the whole of his moral, intellectual and spiritual life by way of the environment of which he forms a natural part." But Weil believed that we had ceased to know the world around us and that "a lot of people think that a little peasant boy of the present day who goes to primary school knows more than Pythagoras did, simply because he can repeat parrot-wise that the earth moves round the sun. In actual fact, he no longer looks up at the heavens." Weil was a teacher, factory worker, member of the French Resistance, and mystic, and she wrote this in the midst of World War II as millions of refugees fled violence and genocide. She warned that uprootedness is the most dangerous malady to which human societies are exposed, leading people to fall into spiritual lethargy or perpetuate their uprootedness against others.

Weil defined rootedness in an interesting way, not as lineage or birthplace but as participation in the life of a community that preserves "certain particular treasures of the past and certain particular expectations for the future." As people grow increasingly distracted from the physical spaces they share with their families, neighbors, and community, retreating from one reality into another, will their sense of uprootedness grow? Virtual worlds might provide us with information, entertainment, and a sense of community, but I doubt whether they could ever fulfill all of our moral, intellectual, and spiritual needs. Increasingly, they seem to pose a threat to consensus or shared expectations for the future.

Interestingly, Nazi party member and philosopher Martin Heidegger warned of many of the same ailments as Simone Weil, in particular that modern society was robbing people of a feeling of being at home in the world. Consequently, Heidegger viewed nostalgia and the longing for home as a condition of modernity. The idea of home is powerful and complex. The philosopher Vincent Vycinas describes it as "an overwhelming, unexchangeable something to which we were subordinate and from which our way of life was oriented and directed." We each bring to this idea of home our own experiences and emotions, of having one or not having one, of deep attachment or pain at losing or being forced from one, feelings that animate us throughout our lives. Even animals follow this pattern of outward journeys and return to a place, on small and grand scales, over a lifetime. Yet it is humans who carry the memories of places we have left behind, to feel the unique longing that in English we call *nostalgia*, from the Greek words *nostos* and *algos*, meaning "return" and "pain."

Johannes Hofer coined the term in the seventeenth century and used it to describe an illness whose symptoms were "persistent thinking of home, melancholia, insomnia, anorexia, loss of thirst, weakness, anxiety, palpitations of the heart, smothering sensations, stupor, and fever." Initially, almost all cases of medically diagnosed nostalgia occurred in Switzerland, where Hofer worked. But no race or nation has a special claim over nostalgia. It is a universal affliction. Within a hundred years after the malady was "discovered," thousands of Scottish soldiers were thought to have died from homesickness. Physicians began documenting cases of nostalgia among Austrian and English soldiers, foreign domestic servants, and African and West Indian enslaved people. In 1897, the psychologist Granville Hall described the possible triggers of nostalgia as "the chirp of crickets, the singing of katydids, the sough of the wind, the pound of the rain, the fragment of a familiar song, and the fleeting resemblance of some place or person to some place or someone in the home situation." Hofer thought that the disease came from within a part of the brain where animal spirits resided, and that when these animals migrated it caused people to be unable to think of anything but home, and could even cause death if untreated. By the early nineteenth century some doctors thought that nostalgia was caused by the blockage of a homing instinct, and others thought it was the result of an exploratory tendency in conflict with the "mother centering tendency" of humans and other animals. Hall called it the conflict between two instincts, an *oikotropic* condition of homesickness that leads us toward home and a *oikifugic* impulse, the desire to travel, that drives one away.

As a kid I moved incessantly from one side of the country to another and back again, moved away from my parents at sixteen, and didn't stop moving until I was in my late twenties. As a result, I have often subscribed to the writer Robyn Davidson's definition of a new kind of nomad in the world: people who are physically but also existentially displaced. "This century has witnessed the greatest upheavals of population in man's history," she writes in her book *Desert Places,* about the Rabari nomads of northwest India. "Yet it is also witnessing the end of traditional nomadism, a description of reality that has been with us since our beginnings— our oldest memory of being. And there are new kinds of nomads, not people who are at home everywhere, but who are at home nowhere. I was one of them."

Likewise, I never felt I had a home to return to. Yet when I really considered the *idea* of home, in Cobb's sense of "a living ecological relationship between an observer and an environment, a person and a place," I found that I did have a reference, and it was the small, scrappy chicken farm that I had loved so much and so briefly as a child.

One day when the purple lilacs were just starting to bloom, I packed a car and headed north with my partner and three-year-old son, toward the past. I made a bet: once we arrived at my old elementary school, I would be able to navigate not only through the entire town but all the way to the farm without asking for directions or using a map. I would get us there as accurately as a homing pigeon using thirty-year-old memories. When we reached the school we walked around the overgrown yard of the boarded-up redbrick building, and I marveled at the size of the maple tree that we used to play kickball under, though it was now a stump. I found hidden corners of the nearby woods and my favorite places to play. And then I got in the driver's seat and drove us without error a few miles down the road.

The trailer we lived in was long gone, but otherwise everywhere I looked I saw evidence of my family's former life there. Pushing back some overgrown brush I found the telephone pole my mother had embedded into her garden; deep within the lilac bush was the flat-topped stone that had provided the centerpiece of my private world. The apple tree branches were craggy and laden with flowers; the chicken coop sagged just like it had before, though now it was filled with lumber. The birch tree I climbed seemed thicker in girth, and the dirt driveway I had to venture to catch the bus in rain or snow was pitifully short, but it was all so close to how I remembered it that I could have walked around blindfolded. We wandered into the freshly mown field and I showed my son the little crick I had swum in. The water still ran clear and strong, its path unchanged by time. I resisted the urge to lie down and never move again.

Revisiting the past is sweet and sad. I had returned but I couldn't go back; time doesn't work that way. Maybe it didn't matter. This place had given me my first ecstatic memories and shaped me like clay. It seemed to me that my responsibility was to re-create those conditions of freedom and belonging for my children, so they would know topophilia and it could guide their way in life. I wanted for them to look around at the immutable topography of the earth or up at the lovely firmament of the sky and recognize their home.

ACKNOWLEDGMENTS

I am indebted to the individuals who shared their experience, perspectives, and heritage with me throughout the course of reporting and researching traditional navigation. They offered their insights with immense kindness, and many reviewed key parts of the manuscript for accuracy in representing historical events and individual stories. My gratitude knows no bounds for their generosity of spirit and time.

In the Arctic, I am particularly grateful to Solomon Awa, John MacDonald, Zacharias Kunuk, Daniel Taukie, Sean Noble-Nowdluk, Matty McNair, Ken MacRury, Myna Ishulutak, Ian Mauro, the wonderful staff of the Nunavut Arctic College Library, Jason Carpenter, and Will Hyndman. A big thank-you to Rick Armstrong and Paul Carolan for hosting me in your homes in Iqaluit. In Australia, I'm thankful to Ray Norris, Margaret Katherine, and the eminent Bill Harney. Thanks to Simon and Phoebe Quilty for putting me up in your beautiful home in Katherine. In Fiji, endless gratitude to Alson Kelen, Peter Nuttall, and the whole village of Korova for their hospitality and insights, as well as Tagi Olosara and his fantastic family in Sigatoka for the *kava* and rugby. In Hawaii, thank you to Kala Baybayan and Timi Gilliom,

Nāʻālehu Anthony, as well as Selena Ching and Sonja Swenson Rogers at the Polynesian Voyaging Society.

At every step of my research a number of anthropologists and scholars gave me the benefit of their years of research and scholarship, including Francesca Merlan, Fred Myers, Claudio Aporta, Thomas Widlok, Joe Genz, Vicente Diaz, David Rubin, Kim Shaw-Williams, Dale Kerwin, Bill Gammage, Tim Ingold, and Harry Heft. Similarly, a group of neuroscientists and researchers guided me through the wonders of the human mind and hippocampus: thank you to Kate Jeffery, Hugo Spiers, Véronique Bohbot, Lynn Nadel, Nora Newcombe, Alessio Travaglia, and Arthur Glenberg. I was touched and inspired by my conversations with Howard Eichenbaum, who sadly passed away in 2017. I will always wonder what more incredible research and ideas about the human brain he might have given us. Special thanks to John Huth for our many conversations over the years and allowing me to visit your lectures at Harvard. Thank you to the participants and organizers of the Royal Institute of Navigation's Animal Navigation conference for allowing me to observe, and in particular Peter Hore and Joe Kirschvink. Also Hugh Dingle for sharing his fascinating insights on animal migration. Kate Compton, thank you for creating Lost Tesla and carefully explaining bots to me.

My research and writing corresponded with a year at the Massachusetts Institute of Technology as a Knight Science Journalism Fellow, and I'm indebted to the incredible staff of the program for their vital support and encouragement. Thank you so much, Deborah Blum, Bettina Urcuioli, David Corcoran, Tom Zeller Jr., and Jane Roberts. Likewise, my fellow fellows were an inspiration: Mark Wolverton, Iván Carrillo, Robert McClure, Fabio Turone, Meera Subramanian, Lauren Whaley, Bianca Toness, Chloe Hecketsweiler, and Rosalia Omungo. At MIT, thank you to the fantastic staff at Hayden and Rotch Libraries, Patrick Winston, Matt Wilson, Wolfgang Victor Hayden Yarlott, and Heather Anne Paxson. At Harvard University, my sincere thanks to James Delbourgo, whose teaching and scholarly perspective was the highlight of my year in Cambridge, as well as Naomi Oreskes and John Stilgoe for their wonderful courses. My sincerest gratitude to Paul Kockelmann at Yale University and Brian Schilder for their incredibly thoughtful review and critical perspective that helped elevate the manuscript, as well as the support of Doron Weber and the Alfred P. Sloan Foundation's Program for the Public Understanding of Science, Technology & Economics that made their review possible.

I am constantly reminded how lucky I am to have my agent, Michelle Tessler, in my corner—thank you for your continued enthusiasm and guidance. And I am extremely grateful to my editor, Elisabeth Dyssegaard, for her friendship and unparalleled talents; thank you for your faith in this endeavor and positive encouragement at every challenging juncture. At St. Martin's Press I'm very fortunate to have Alan Bradshaw and Laura Apperson helping me through the process with kindness and patience. Special thanks to Emma Piper-Burkett for many heartening conversations and Tom Peter for your friendship and getting lost on a bike in London with me, thereby proving the thesis of my book in real time. Kristi Lutz and Pi Waller, thank you for lending me your lovely home in Long Island for much-needed writing time. Also Elliott Prasse-Freeman for help at a crucial moment and the carton of James Scott's chicken eggs.

Thank you to my family for your cheerleading: Chris Miller, Mark Miller, Ciaran O'Connor, Jane O'Connor George, and Margaret Parker, and the support of my fantastic and loving parents, Rory O'Connor and Katherine Miller. I haven't got the words to adequately thank my grandparents Bob and Janet Miller for their unceasing support—I love you dearly.

And last, my profound thanks to Bryan Parker for making it all possible. Your infinite equanimity and humor make you the best travel companion anyone could wish for in life.

NOTES

Prologue

2   "allow the fact that" Audrey Niffenegger, *Her Fearful Symmetry: A Novel*, 2009, 264.

4   "travels today to the Arctic" April White, "The Intrepid '20s Women Who Formed an All-Female Global Exploration Society," *Atlas Obscura*, April 12, 2017, http://www.atlasobscura.com/articles/society-of-woman-geographers.

4   "after three thousand years" Marshall McLuhan, quoted in Lewis H. Lapham, *Understanding Media: The Extensions of Man*, reprint edition, 1994, 3.

4   "represent[ed] a substantial narrowing" James C. Scott, *Against the Grain: A Deep History of the Earliest States*, first edition, 2017, 91.

5   "For thousands of years" A. Ardila, "Historical Evolution of Spatial Abilities," *Behavioural Neurology* 6, no. 2 (1993): 83–87, https://doi.org/10.3233/BEN-1993-6203.

5   "the old ways" Barbara Moran, "The Joy of Driving without GPS," *Boston Globe*, August 8, 2017, https://www.bostonglobe.com/magazine/2017/08/08/the-joy-driving -without-gps/W36dJaTGw05YFdzyixhj3M/story.html.

8   "The idea is that during" Matthew Wilson, presentation at Massachusetts Institute of Technology, December 5, 2016.

8   "the hippocampus is a phylogenetically old" Eleanor A. Maguire et al., "Navigation-Related Structural Change in the Hippocampi of Taxi Drivers," *Proceedings of the National Academy of Sciences* 97, no. 8 (April 11, 2000): 4398–403, doi.org/10.1073 /pnas.070039597.

9   "If navigation was the primary purpose" Howard Eichenbaum, Interview with author, October 18, 2016.

13   "With nature as your" Harold Gatty and J. H. Doolittle, *Nature Is Your Guide: How to Find Your Way on Land and Sea by Observing Nature*, 1979, back flap.

13   "For Nenets, navigating is" Andrei Golovnev, quoted in Kirill V. Istomin and Mark J. Dwyer, "Finding the Way: A Critical Discussion of Anthropological Theories of Human Spatial Orientation with Reference to Reindeer Herders of Northeastern Europe and Western Siberia," *Current Anthropology* 50, no. 1 (February 1, 2009): 29–49, doi.org/10.1086/595624.

15   "They wouldn't get lost" Ken MacRury, Interview with author, June 14, 2016.

15 "You drive out on the weekend" Thomas Widlok, Interview with author, July 26, 2016.

16 "to see the world" Robyn Davidson, *Desert Places*, first edition, 1996, 146.

16 "the ability to determine a route" Reginald Golledge, quoted in Dario Guiducci and Ariane Burke, "Reading the Landscape: Legible Environments and Hominin Dispersals," *Evolutionary Anthropology* 25, no. 3 (May 6, 2016): 133–41, doi.org/10.1002/evan.21484.

17 "We are told that vision" James J. Gibson, *The Ecological Approach to Visual Perception: Classic Edition*, first edition, 2014, xiii.

17 "all persons who want" James J. Gibson, *The Senses Considered as Perceptual Systems*, revised edition, 1983, 321.

19 "It's only a slight" Harry Heft, Interview with author, March 14, 2017.

19 "By using a GPS" Tristan Gooley, Email with author, May 11, 2015.

19 "We want to go faster" Tim Ingold, Interview with author, March 30, 2016.

The Last Roadless Place

24 "poor disciple" James McDermott, *Martin Frobisher: Elizabethan Privateer*, 2001, 133.

25 "When adventure does not come" Jean Malaurie, *The Last Kings of Thule: A Year Among the Polar Eskimos of Greenland*, 1956, 202.

25 "so-called advanced" Jean Malaurie, *Ultima Thulé: Explorers and Natives of the Polar North*, 2003, 146.

26 "extremely high mobility" Max Friesen, "North America: Paleoeskimo and Inuit Archaeology," in *Encyclopedia of Global Human Migration*, ed. by Immanuel Ness, 2013, https://www.academia.edu/5314092/North_America_Paleoeskimo_and_Inuit_archaeology_Encyclopedia_of_Global_Human_Migration_.

26 "Even burial grounds" Leo Ussak, quoted in Milton Freeman Research Limited, *Inuit Land Use and Occupancy Project: A Report*, 1976, 192.

28 "The land does not change" Scott Brachmayer, *Kajutaijuq*, film short, 2015, http://www.imdb.com/title/tt3826696/.

30 "I was born in a sod house" Solomon Awa, Interview with author, May 10, 2015.

33 "They had to" Nancy Wachowich, Apphia Agalakti Awa, Rhoda Kaukjak Katsak, and Sandra Pikujak Katsak, *Saqiyuq: Stories from the Lives of Three Inuit Women*, 2001, 106.

33 "My husband asked" Ibid., 108.

34 "the color of sky" Alfred K. Siewers, "Colors of the Winds, Landscapes of Creation," in *Strange Beauty*, The New Middle Ages, 2009, 97.

Memoryscapes

38 "astonishing precision" Robert A. Rundstrom, "A Cultural Interpretation of Inuit Map Accuracy," *Geographical Review* 80, no. 2 (1990): 155–68, //doi.org/10.2307/215479.

38 "All the places where" Kenn Harper, "Wooden Maps," *Nunatsiaq Online*, April 11, 2014, http://www.nunatsiaq.com/stories/article/65674taissumani_april_11/.

38 "The historical record" Rundstrom, "A Cultural Interpretation of Inuit Map Accuracy."

39 "these people sail" Ben Finney, *Voyage of Rediscovery: A Cultural Odyssey through Polynesia*, 1994, 11.

39 "a non-literate man" Margarette Lincoln, ed., *Science and Exploration in the Pacific: European Voyages to the Southern Oceans in the Eighteenth Century*, 1998, 127.

40 "non-industrialists" R. Robin Baker, *Human Navigation and the Sixth Sense*, first edition, 1981, 48.

40 "In the flat country" Harold Gatty and J. H. Doolittle, *Nature Is Your Guide: How to Find Your Way on Land and Sea by Observing Nature*, 1979, 48.

40   "instinctive sense, beyond our comprehension" Phillip Lionel Barton, "Maori Cartography and the European Encounter," in *The History of Cartography: Cartography in the Traditional African, American, Arctic, Australian, and Pacific Societies*, ed. by David Woodward and G. Malcolm Lewis, first edition, vol. 2, book 3, 1998, 496.

40   "guided by a kind" Baker, *Human Navigation and the Sixth Sense*, 47.

41   "preserved useful variations" Charles Darwin, "Origin of Certain Instincts," *Nature* 7, no. 179 (April 3, 1873): 007417a0, doi.org/10.1038/007417a0.

41   "intricate labyrinths of ice" Ferdinand Petrovich Baron Wrangel, *Narrative of an Expedition to the Polar Sea, in the Years 1820, 1821, 1822 & 1823 Commanded by Lieutenant, Now Admiral Ferdinand Von Wrangel*, ed. by Edward Sabine, 1841, 40.

42   "Throughout most of human evolution" Baker, *Human Navigation and the Sixth Sense*, 8.

42   "In our Western civilization" Harold Gatty, *Finding Your Way Without Map or Compass*, reprint edition, 1999, 30.

43   "Eskimo is a scientist" Richard K. Nelson, *Hunters of the Northern Ice*, 1972, xxii.

44   "There is no" Ibid., xxiii.

44   "shallow stream valleys" Ibid., 102.

45   "How is it that" Claudio Aporta, "Inuit Orienting: Traveling along Familiar Horizons," *Sensory Studies*, n.d., http://www.sensorystudies.org/inuit-orienting-traveling-along-familiar-horizons/, accessed February 27, 2015.

45   "One of the main differences" Claudio Aporta, "Old Routes, New Trails: Contemporary Inuit Travel and Orienting in Igloolik, Nunavut," thesis, 2003, 82.

46   "through named features" Ibid., 3.

46   "The relationships between" J. M. Fladmark and Thor Heyerdahl, *Heritage and Identity: Shaping the Nations of the North*, 2015, 231.

48   "Thinking at the highest level" Charles O. Frake, "Cognitive Maps of Time and Tide among Medieval Seafarers," *Man* 20, no. 2 (1985): 254–70, doi.org/10.2307/2802384.

49   "honey-tongued" Frances Yates, *The Art of Memory*, 2014, 27.

50   "at the great nerve" Ibid., 127.

50   "particulars out of the mass" Ibid., 372.

51   "For this invention will produce" Ibid., 38.

51   "superior memorizers" Eleanor A. Maguire, Elizabeth R. Valentine, John M. Wilding, and Narinder Kapur, "Routes to Remembering: The Brains behind Superior Memory," *Nature Neuroscience* 6, no. 1 (January 2003): 90–95, doi.org/10.1038/nn988.

52   "journey" Gordon H. Bower, "Analysis of a Mnemonic Device: Modern Psychology Uncovers the Powerful Components of an Ancient System for Improving Memory," *American Scientist* 58, no. 5 (1970): 496–510.

53   "The results from the present study" Eleanor A. Maguire, Katherine Woollett, and Hugo J. Spiers, "London Taxi Drivers and Bus Drivers: A Structural MRI and Neuropsychological Analysis," *Hippocampus* 16, no. 12 (December 1, 2006): 1091–101, doi.org/10.1002/hipo.20233.

54   "evolution of our civilisation" Gatty, *Finding Your Way Without Map or Compass*, 53.

54   "trail-blazing pioneer" Tristan Gooley, "The Navigator That Time Lost," The Natural Navigator, 2009, https://www.naturalnavigator.com/the-library/the-navigator-that-time-lost.

55   "natural navigators" Gatty and Doolittle, *Nature Is Your Guide*, 219.

55   "[Scientists] build a wall" Ibid., 27.

Why Children Are Amnesiacs

58   "Why would nature" Elizabeth Marozzi and Kathryn J. Jeffery, "Place, Space and Memory Cells," *Current Biology* 22, no. 22 (2012): R939–42.

58   "I think the field" Kate Jeffery, Interview with author, April 14, 2016.

60   "Hitherto it has not" Sigmund Freud, *Freud on Women: A Reader*, 1992, 106.

61 "[They believed] that the advent" Nora Newcombe, Interview with author, July 29, 2016.

62 "This emergence of" Moshe Bar, ed., *Predictions in the Brain: Using Our Past to Generate a Future*, revised edition, 2011, 351.

63 "like waking from a dream" Larry R. Squire, "The Legacy of Patient H.M. for Neuroscience," *Neuron* 61, no. 1 (January 15, 2009): 6–9, https://doi.org/10.1016/j.neuron.2008.12.023.

64 "The biggest surge in synaptic" John O'Keefe and Lynn Nadel, *The Hippocampus as a Cognitive Map*, 1978, 114.

65 "does not have a representation" Ibid., 241.

66 "This pattern suggests" Lynn Nadel and Stuart Zola-Morgan, "Infantile Amnesia: A Neurobiological Perspective," in *Infant Memory: Its Relation to Normal and Pathological Memory in Humans and Other Animals*, vol. 9, ed. by Morris Moscovitch, 2012, 145.

66 "The hippocampus, it's not" Lynn Nadel, Interview with author, July 29, 2016.

69 "Critical periods are when" Alessio Travaglia, Interview with author, August 2, 2016.

75 "All living beings" Brett Buchanan, *Onto-Ethologies: The Animal Environments of Uexküll, Heidegger, Merleau-Ponty, and Deleuze*, 2009, 26.

Birds, Bees, Wolves, and Whales

78 "Inuit are pragmatists" Ken MacRury, Interview with author, June 14, 2016.

78 "The faster you traverse" John MacDonald, Interview with author, January 14, 2016.

79 "The term can be translated as" John MacDonald, Email with author, November 16, 2017.

79 "one who moves away" William A. Lovis and Robert Whallon, *Marking the Land: Hunter-Gatherer Creation of Meaning in Their Environment*, 2016, 85.

80 "Men and wolves" Roger Peters, "Cognitive Maps in Wolves and Men," *Environmental Design Research* 2 (1973): 247–53.

80 "For wolves, the reality" Ibid.

83 "spatial primitives" Kate J. Jeffery et al., "Animal Navigation—A Synthesis," in *Animal Thinking: Contemporary Issues in Comparative Cognition*, ed. by Julia Fischer and Randolf Menzel, 2011, 59.

84 "eerily consistent" James L. Gould and Carol Grant Gould, *Nature's Compass: The Mystery of Animal Navigation*, 2012, 38.

85 "preprogrammed" Hugh Dingle, *Migration: The Biology of Life on the Move*, first edition, 1996, 214.

89 "as straight as an arrow" Travis W. Horton et al., "Straight as an Arrow: Humpback Whales Swim Constant Course Tracks during Long-Distance Migration," *Biology Letter* (April 20, 2011): rsbl20110279, doi.org/10.1098/rsbl.2011.0279.

91 "maddeningly difficult" "The Magnetic Sense Is More Complex Than Iron Bits," *Evolution News*, April 29, 2016, https://evolutionnews.org/2016/04/the_magnetic_se/.

92 "the deep substrate" Matthew Cobb, "Are We Ready for Quantum Biology?" *New Scientist*, November 12, 2014, https://www.newscientist.com/article/mg22429950-700-are-we-ready-for-quantum-biology/.

94 "chemical compass" Klaus Schulten, Charles E. Swenberg, and Albert Weller, "A Biomagnetic Sensory Mechanism Based on Magnetic Field Modulated Coherent Electron Spin Motion," *Zeitschrift für Physikalische Chemie* 111, no. 1 (1978): 1–5, doi.org/10.1524/zpch.1978.111.1.001.

96 "migratory syndromes" Dingle, *Migration: The Biology of Life on the Move*, 33.

96 "hidden drive at the right time" Peter Berthold, *Bird Migration: A General Survey*, 2001, 11.

96 "pinioned wild goose" Charles Darwin, *Charles Darwin's Shorter Publications, 1829–1883*, 2009, 380.

97 "How do these geese" Elisabeth Kübler-Ross, *The Wheel of Life*, 2012, 106.

Navigation Made Us Human

98 "that which acts" Norman Hallendy, *Inuksuit: Silent Messengers of the Arctic*, first trade paper edition, 2001, 46.

100 "How did foragers" Daniel Casasanto, "Space for Thinking," in *Language, Cognition and Space*, ed. Vyvyan Evans and Paul Chilton, 2010, 455.

101 "Man has been a hunter" Carlo Ginzburg, *Clues, Myths, and the Historical Method*, trans. John Tedeschi and Anne C. Tedeschi, reprint edition, 2013, 102.

102 "The hunter would have" Ibid., 103.

102 "[T]he notion that the world" Derek Bickerton, *More Than Nature Needs: Language, Mind, and Evolution*, 2014, 88.

103 "lethal parody of the" Alfred Gell, "How to Read a Map: Remarks on the Practical Logic of Navigation," *Man* 20, no. 2 (June 1985): 27–86, https://doi.org/10.2307/2802385.

103 "We read in it" Ibid., 26.

105 "I have to mentally" Kim Shaw-Williams, "The Triggering Track-Ways Theory," thesis, 2011, http://researcharchive.vuw.ac.nz/handle/10063/1967.

106 "Although common sense endows" Endel Tulving, "Episodic Memory and Common Sense: How Far Apart?" *Philosophical Transactions of the Royal Society of London. Series B, Biological Sciences* 356, no. 1413 (September 29, 2001): 1505–15, doi.org/10.1098/rstb.2001.0937.

106 "Lest someone worry" Ibid.

110 "deconfuser" Ibid.

110 "the faculty of visualization" Norman Hellendy, "Tukiliit: The Stone People Who Live in the Wind; An Introduction to Inuksuit and Other Stone Figures of the North," January 18, 2017, https://docslide.com.br/documents/tukiliit-the-stone-people-who-live-in-the-wind-an-introduction-to-inuksuit.html.

112 "Right now hunters go" William Hyndman, Interview with author, May 2, 2016.

114 "shaped like an animal's heart" "Introduction: Place Names in Nunarat," Inuit Heritage Trust: Place Names Program, http://ihti.ca/eng/place-names/pn-index.html?agree=0.

114 "Sometimes we name them" John Bennett and Susan Rowley, *Uqalurait: An Oral History of Nunavut*, 2004, 113.

The Storytelling Computer

118 "Narrative serves as a vehicle" Lawrence S. Sugiyama and Michelle Scalise Sugiyama, "Humanized Topography: Storytelling as a Wayfinding Strategy," *American Anthropologist* 5, http://pages.uoregon.edu/sugiyama/docs/StoryMapsMainDocument[1].pdf.

119 "culture hero" James Alexander Teit, *Traditions of the Thompson River Indians of British Columbia*, 1898, 7.

119 "All being seated" Richard Irving Dodge and General William Tecumseh Sherman, *Our Wild Indians: Thirty-Three Years' Personal Experience among the Red Men of the Great West—A Popular Account of Their Social Life, Religion, Habits, Traits, Customs, Exploits, Etc.*, reprint edition, 1978, 552.

120 "the Pawnees had a" Gene Weltfish, *The Lost Universe: Pawnee Life and Culture*, 1977, 172.

120 "talked names" Keith H. Basso, *Wisdom Sits in Places: Landscape and Language among the Western Apache*, 1996, 45–47.

121 "Your life is like a trail" Ibid., 126.

123 "Merge" Robert C. Berwick and Noam Chomsky, *Why Only Us: Language and Evolution*, 2016, 10.

123 "We can construct these elaborate" Robert C. Berwick, "Why Only Us," Classroom 10-250, Massachusetts Institute of Technology, November 28, 2016.

128 "Stories are how" Wolfgang Yarlott and Victor Hayden, "Old Man Coyote Stories: Cross-Cultural Story Understanding in the Genesis Story Understanding System," thesis, 2014, http://dspace.mit.edu/handle/1721.1/91880.

128 "How he did this" Ibid.

128 "Start description of" Ibid., 38.

129 "Start experiment" Ibid., 80.

130 "I believe this is a" Ibid., 60.

Supernomads

136 "Well some other tribes" Canning Stock Route Project, http://mira.canningstockroute project.com/.

136 "charge into the spatial" Dale Kerwin, *Aboriginal Dreaming Paths and Trading Routes: The Colonisation of the Australian Economic Landscape*, 2010, 159.

137 "the whole Western desert" David Lewis, "Observations on Route Finding and Spatial Orientation among the Aboriginal Peoples of the Western Desert Region of Central Australia," *Oceania* 46, no. 4 (1976): 249–82.

137 "try and get him" "Putuparri Tom Lawford: Oral History," Canning Stock Route Project, http://mira.canningstockrouteproject.com/node/3060, accessed February 10, 2016.

138 "The earth is the repository" Deborah Bird Rose, *Dingo Makes Us Human: Life and Land in an Australian Aboriginal Culture*, 2000, 57.

138 "There are wells" "Putuparri Tom Lawford: Oral History."

138 "everywhen" D. J. Mulvaney, "Stanner, William Edward (Bill) (1905–1981)," in *Australian Dictionary of Biography*, n.d., http://adb.anu.edu.au/biography/stanner -william-edward-bill-15541.

138 "theory of everything" Robyn Davidson, *Quarterly Essay 24: No Fixed Address: Nomads and the Fate of the Planet*, 2006, 13.

139 "a spiritual realm" Ibid., 14.

139 "Thus the landscape" David Turnbull and Helen Watson, *Maps Are Territories: Science Is an Atlas; A Portfolio of Exhibits*, 1989, 30.

140 "super-nomads" Scott Cane, *First Footprints: The Epic Story of the First Australians*, main edition, 2014, 30.

141 "follows behind us" Rose, *Dingo Makes Us Human*, 2000, 205.

141 "We are linked by song" Dale Kerwin, *Aboriginal Dreaming Paths and Trading Routes: The Colonisation of the Australian Economic Landscape*, 2010, 49.

142 "resides in mnemonics" Ibid., 83.

142 "[The ancestors] know" Ibid., 37.

143 "deeper significance" Alfred Reginald Radcliffe-Brown, Raymond William Firth, and Adolphus Peter Elkin, *Oceania* (1975): 271.

143 "some of the world's earliest" Patrick D. Nunn and Nicholas J. Reid, "Aboriginal Memories of Inundation of the Australian Coast Dating from More Than 7000 Years Ago," *Australian Geographer* 47, no 1 (September 7, 2015): 47, doi/abs/10.1080/0004 9182.2015.1077539.

143 "For example, a man" Ibid.

144 "There are no one-scene epics" David C. Rubin, *Memory in Oral Traditions: The Cognitive Psychology of Epic, Ballads, and Counting-out Rhymes*, 1997, 62.

145 "only in its performance" Ibid., 114.

Dreamtime Cartography

146 "vast howling wilderness" "Yiwarra Kuju," National Museum of Australia, http://www.nma.gov.au/education/resources/units_of_work/yiwarra_kuju.

146 "Where, we ask, is" Bill Gammage, *The Biggest Estate on Earth: How Aborigines Made Australia*, reprint edition, 2013, 309.

147 "Whitefellas just reckon go" "Aboriginal Guides," Canning Stock Route Project, http://www.canningstockrouteproject.com/history/story-aboriginal-guides/.

148 "While there are" David Lewis, "Route Finding and Spatial Orientation," *Oceania* 46, no. 4 (1975).

149 "A single visit" David Lewis, "Route Finding by Desert Aborigines in Australia," *Journal of Navigation* 29, no. 1 (January 1976): 21–38, doi.org/10.1017/S0373463300043307.

149 "I have a feeling" Ibid.

149 "How do you know" Lewis, "Route Finding and Spatial Orientation," 262.

150 "Pintupi's route-finding" Ibid.

150 "All my preconceived ideas" Ibid.

151 "Often it is difficult" Peter Sutton, "Aboriginal Maps and Plans," in *The History of Cartography: Cartography in the Traditional African, American, Arctic, Australian, and Pacific Societies*, ed. David Woodward and G. Malcolm Lewis, first ed., vol. 2, 1998, 407.

151 "radical in its fundamental implication" Philip G. Jones, "Norman B. Tindale Obituary," December 1995, https://www.anu.edu.au/linguistics/nash/aust/nbt/obituary.html.

152 "A lot of kids run" Hetti Perkins, *Art Plus Soul*, 2010, 58.

152 "I was a boy" Ibid.

154 "I failed fully to understand" Lewis, "Route Finding and Spatial Orientation," 249–82.

155 "Aboriginal roads and tracks" Dale Kerwin, *Aboriginal Dreaming Paths and Trading Routes: The Colonisation of the Australian Economic Landscape*, 2010, 114.

155 "Like western topographic maps" Ibid., 47.

155 "European maps are not" David Turnbull and Helen Watson, *Maps Are Territories: Science Is an Atlas; A Portfolio of Exhibits*, 1989, 51.

156 "It is a stark country" Fred Myers, *Pintupi Country, Pintupi Self: Sentiment, Place, and Politics among Western Desert Aborigines*, 1991, 11.

159 "undulations hardly deserving" David Lewis, Curriculum Development Centre, and Aboriginal Arts Board, "The Way of the Nomad," in *From Earlier Fleets: Hemisphere—An Aboriginal Anthology*, ed. Kenneth Russell Henderson, 1978.

160 "some kind of *dynamic*" Lewis, "Route Finding and Spatial Orientation," 262.

Space and Time in the Brain

161 "Over a period of months" John O'Keefe, "Biographical," The Nobel Foundation, 2014, https://www.nobelprize.org/nobel_prizes/medicine/laureates/2014/okeefe-bio.html.

162 "In thinking about these results" Ibid.

164 "Everything important in" James C. Goodwin, "A-Mazing Research," *American Psychological Association* 43, no. 2 (February 2012), http://www.apa.org/monitor/2012/02/research.aspx.

164 "telephone switchboard" Edward C. Tolman, "Cognitive Maps in Rats and Men," *Psychological Review* 55, no. 4 (July 1948): 189–208, http://dx.doi.org/10.1037/h0061626.

165 "cognitive-like map" Ibid.

165 "brief, cavalier, and dogmatic" Ibid.

165 "My only answer is" Ibid.

166 "who first dreamed of" John O'Keefe and Lynn Nadel, *The Hippocampus as a Cognitive Map*, 1978.

167 "Space plays a role" Ibid., 6.

167 "Constructor of Brains" Ibid.

167 "ineliminable property of our" Lynn Nadel, "The Hippocampus and Space Revisited," *Hippocampus* 1, no. 3 (July 3, 1991): 221–29, doi.org/10.1002/hipo.450010302.

168 "Space was a *way*" O'Keefe and Nadel, *The Hippocampus as a Cognitive Map*, 19.

168 "hedged his bets" Ibid., 6, 296.

170 "The idea was that" O'Keefe, "Biographical."

171 "combined into supra modal" Elizabeth Marozzi and Kathryn J. Jeffery, "Place, Space and Memory Cells," *Current Biology* 22, no. 22 (2012): R939–42.

172 "If you damage the" Matthew Wilson, Presentation at Massachusetts Institute of Technology, December 5, 2016.

176 "temporarily structured experiences" Daniela Schiller et al., "Memory and Space: Towards an Understanding of the Cognitive Map," *Journal of Neuroscience* 35, no. 41 (October 14, 2015): 13904–11, doi.org/10.1523/JNEUROSCI.2618-15.2015.

177 "The hippocampal system" Howard Eichenbaum and Neal J. Cohen, "Can We Reconcile the Declarative Memory and Spatial Navigation Views on Hippocampal Function?" *Neuron* 83, no. 4 (August 20, 2014): 764–70, doi.org/10.1016/j.neuron.2014.07.032.

178 "The cognitive map is" Hugo J. Spiers, Interview with author, April 12, 2016.

179 "When I think about" Harry Heft, Interview with author, March 16, 2017.

181 "Wayfinding to a specific destination" Harry Heft, "The Ecological Approach to Navigation: A Gibsonian Perspective," in *The Construction of Cognitive Maps*, ed. J. Portugali, 1996, 105–32, doi.org/10.1007/978-0-585-33485-1_6.

Among the Lightning People

183 "We talk about emus" Ray P. Norris and Bill Yidumduma Harney, "Songlines and Navigation in Wardaman and Other Australian Aboriginal Cultures," *Journal of Astronomical History and Heritage* 17, no. 2 (April 9, 2014), http://arxiv.org/abs/1404.2361.

185 "Dead too are the" William Edward Harney, *Life among the Aborigines*, 1957, 38.

185 "If you lay on" Norris and Harney, "Songlines and Navigation in Wardaman and Other Australian Aboriginal Cultures."

190 "as intimately as humans" Bill Gammage, *The Biggest Estate on Earth: How Aborigines Made Australia*, reprint edition, 2013, 122.

190 "In its notions of time" Ibid., 132.

192 "constantly impressed by how" Isabel McBryde, "Travellers in Storied Landscapes: A Case Study in Exchanges and Heritage," *Aboriginal History* 24 (2000): 152–74.

193 "Well, everyone else is" Jan Wositzky, *Born under the Paperbark Tree: A Man's Life*, ed. Yidumduma Bill Harney, revised edition, 1998, 178.

194 "Yeah, that's my proper" Ibid., 179.

195 "I grew up with" Hugh Cairns, *Dark Sparklers: Yidumduma's Wardaman Aboriginal Astronomy Northern Australia 2003*, 2003, 16.

You Say Left, I Say North

201 "Probably the apprehension" Zoltan Kovecses, *Language, Mind, and Culture: A Practical Introduction*, 2006, 13.

201 "[J]ust as we think" Stephen C. Levinson, *Space in Language and Cognition: Explorations in Cognitive Diversity*, 2003, 114.

202 "You always know which" Ibid., 131.

202 "Speakers of languages" Ibid., 21.

203 "Nothing like this" Ibid., 127.

204 "partaking of nature" Thomas Widlok, "The Social Relationships of Changing Hai‖om Hunter Gatherers in Northern Namibia, 1990–1994," 1994, 210.

207 "Language" Asifa Majid, Melissa Bowerman, Sotaro Kita, Daniel B. M. Haun, and

Stephen C. Levinson, "Can Language Restructure Cognition? The Case for Space," *Trends in Cognitive Sciences* 8, no. 3 (March 2004): 108–14, doi.org/10.1016/j.tics .2004.01.003.

208 "prolonged social interaction" Thomas Widlok, "Orientation in the Wild: The Shared Cognition of Hai‖om Bushpeople," *Journal of the Royal Anthropological Institute* 3, no. 2 (1997): 317–32, https://doi.org/10.2307/3035022.

209 "patchwork of landscape" Ibid.

210 "The central point" Kirill V. Istomin and Mark J. Dwyer, "Finding the Way: A Critical Discussion of Anthropological Theories of Human Spatial Orientation with Reference to Reindeer Herders of Northeastern Europe and Western Siberia," *Current Anthropology* 50, no. 1 (February 1, 2009): 29–49, doi.org/10.1086/595624.

210 "a skilled performance" Tim Ingold, *The Perception of the Environment: Essays on Livelihood, Dwelling and Skill*, first edition, 2011, 220.

211 "not in but along" Nuccio Mazzullo and Tim Ingold, "Being Along: Place, Time and Movement among Sámi People," in *Mobility and Place: Enacting Northern European Peripheries*, ed. Jørgen Ole Bærenholdt and Brynhild Granås, 2012.

212 "tantamount to the organism's" Ingold, *The Perception of the Environment: Essays on Livelihood, Dwelling and Skill*, 3.

212 "knowing as we go" Ibid., 229.

213 "[I]ndeed he may have" Ibid., 234.

213 "the path, like the musical" Ibid., 238.

213 "journeys through space" Howard Eichenbaum, "Hippocampus: Mapping or Memory?" *Current Biology* 10, no. 21 (November 1, 2000): R785–87, doi.org/10.1016 /S0960-9822(00)00763-6.

Empiricism at Harvard

217 "oldest science" Louis Liebenberg, *The Art of Tracking: The Origin of Science*, first edition, 2012, xv.

218 "have to create a working" Ibid., 116.

219 "The modern scientist" Ibid., 57.

219 "at least some of the first" Louis Liebenberg, *The Origin of Science: On the Evolutionary Roots of Science and Its Implications for Self-Education and Citizen Science*, 2013, 17.

223 "Sailors today have no need" Charles O. Frake, "Cognitive Maps of Time and Tide among Medieval Seafarers," *Man* 20, no. 2 (1985): 254–70, https://doi.org/10.2307 /2802384.

226 "spatial field of multitudes" "Wave Piloting in the Marshall Islands," Conference, Radcliffe Institute for Advanced Study, Harvard University, June 20, 2017.

227 "so vast that the human" J. C. Beaglehole, *The Life of Captain James Cook*, 1992, 109.

227 "[S]tandard histories of cartography" Ben Finney, "Nautical Cartography and Traditional Navigation in Oceania," in *The History of Cartography: Cartography in the Traditional African, American, Arctic, Australian, and Pacific Societies*, ed. David Woodward and G. Malcolm Lewis, first edition, vol. 2, book 3, 1998, 444.

228 "All these things" Nāʻālehu Anthony, presentation at "The *Hōkūleʻa*: Indigenous Resurgence from Hawaiʻi to Mannahatta," New York University, March 31, 2016.

230 "breaking open the turtle" Joseph Genz, "Navigating the Revival of Voyaging in the Marshall Islands: Predicaments of Preservation and Possibilities of Collaboration," *Contemporary Pacific* 23, no. 1 (March 26, 2011): 1–34, doi.org/10.1353/cp.2011.0017.

230 "The physical and social" Ibid.

230 "When I saw the bomb's light" Ibid.

232 "from our scientific" Joseph Genz, Jerome Aucan, Mark Merrifield, Ben Finney, Korent Joel, and Alson Kelen, "Wave Navigation in the Marshall Islands: Comparing Indigenous and Western Scientific Knowledge of the Ocean," *Oceanography* 22, no. 2 (2009): 234–45.

Astronauts of Oceania

236 "If you want to cause change" Nā'ālehu Anthony, presentation at "The *Hōkūle'a*: Indigenous Resurgence from Hawai'i to Mannahatta," New York University, March 31, 2016.

236 "The welcoming of the" Vicente Diaz, presentation at "The *Hōkūle'a*: Indigenous Resurgence from Hawai'i to Mannahatta," New York University, March 31, 2016.

237 "science and technology" Vicente Diaz, "Lost in Translation and Found in Constipation: Unstopping the Flow of Intangible Cultural Heritage with the Embodied Tangibilities of Traditional Carolinian Seafaring Culture," International Symposium on Negotiating Intangible Cultural Heritage, National Ethnology Museum, Osaka, Japan, 2017.

238 "the process of thinking" Thomas Gladwin, *East Is a Big Bird: Navigation and Logic on Puluwat Atoll*, 1995, preface.

238 "almost every young man" Ibid., 37.

238 "The voyages are often" Ibid., 42.

238 "reopened" David Lewis, "Memory and Intelligence in Navigation: Review of *East Is a Big Bird*. Gladwin Thomas. Harvard University Press," *Journal of Navigation* 24, no. 3 (July 1971): 423–24, doi.org/10.1017/S0373463300048426.

238 "With such abounding" Gladwin, *East Is a Big Bird: Navigation and Logic on Puluwat Atoll*, 37.

239 "[W]hen set to tune" Vicente Diaz, "No Island Is an Island," in *Native Studies Keywords*, ed. by Stephanie Nohelani Teves, Andrea Smith, and Michelle Raheja, 2015, 90–107.

239 "This is a body of knowledge" Ibid., 131.

240 "This picture he uses" Ibid., 182.

241 "the canoe was the center" Sam Low, *Hawaiki Rising: Hōkūle'a, Nainoa Thompson, and the Hawaiian Renaissance*, 2013, 61.

242 "I can't explain it" Ibid., 277.

243 "house of instruction" Kala Baybayan, Interview with author, January 12, 2017.

244 "my *'aumakua*" Ibid., 322.

Navigating Climate Change

249 "Earlier, the hunters" Chris Mooney, "In Greenland's Northernmost Village, a Melting Arctic Threatens the Age-Old Hunt," *Washington Post*, April 29, 2017, https://www.washingtonpost.com/business/economy/in-greenlands-northernmost-village-a-melting-arctic-threatens-the-age-old-hunt/2017/04/29/764ba9be-1bb3-11e7-bcc2-7d1a0973e7b2_story.html.

250 "Tuvaluans will not cease" Robert A. McLeman, *Climate and Human Migration: Past Experiences, Future Challenges*, first edition, 2013, 199.

252 "Current transport, especially domestic" Peter Roger Nuttall, "Sailing for Sustainability: The Potential of Sail Technology as an Adaptation Tool for Oceania; A Voyage of Inquiry and Interrogation through the Lens of a Fijian Case Study," 2013, 15.

252 "the connection, the interface" Ibid., 46.

253 "All indicators were" Ibid., 35.

254 "My young sons" Ibid., 36.

256 "far superior to those" Peter Nuttall, Paul D'Arcy, and Colin Philp, "Waqa Tabu—Sacred Ships: The Fijian Drua," *International Journal of Maritime History* 26, no. 3 (August 1, 2014): 427–50, doi.org/10.1177/0843871414542736.

260 "Knut, next to me" Ibid.

This Is Your Brain on GPS

262 "I think [spatial ability]" Tom Clynes, "How to Raise a Genius: Lessons from a 45-Year Study of Super-Smart Children," *Nature News* 537, no. 7619 (September 8, 2016): 152, doi.org/10.1038/537152a.

262 "People who have shrunk" Véronique Bohbot, Interview with author, June 30, 2016.

268 "Global Determinants" Antoine Coutrot et al., "Global Determinants of Navigation Ability," *bioRxiv* (September 18, 2017): 188870, doi.org/10.1101/188870.

270 "From a dead-reckoning" Stephen C. Levinson, *Space in Language and Cognition: Explorations in Cognitive Diversity*, 2003, 238.

271 "There are a lot of researchers" Hugo J. Spiers, Interview with author, April 12, 2016.

271 "Our results fit" "Satnavs 'Switch Off' Parts of the Brain," University College London News, March 21, 2017, http://www.ucl.ac.uk/news/news-articles/0317/210317 -satnav-brain-hippocampus.

273 "The GPS diffuses that" Harry Heft, Interview with author, March 16, 2017.

274 "Whatever is beneficial" Sinéad L. Mullally and Eleanor A. Maguire, "Memory, Imagination, and Predicting the Future," *Neuroscientist* 20, no. 3 (June 2014): 220– 34, doi.org/10.1177/1073858413495091.

275 "likely to be useful" Benjamin Baird, Jonathan Smallwood, and Jonathan W. Schooler, "Back to the Future: Autobiographical Planning and the Functionality of Mind-Wandering," *Consciousness and Cognition, From Dreams to Psychosis* 20, no. 4 (December 1, 2011): 1604–11, doi.org/10.1016/j.concog.2011.08.007.

Lost Tesla

277 "Be Anywhere" Max Londberg, "KC to STL in 20 Minutes? System That Could Threaten Speed of Sound May Come to Missouri," *Kansas City Star*, April 7, 2017, http://www.kansascity.com/news/local/article143315884.html.

278 "Whatever our idea of" Mark Vanhoenacker, *Skyfaring: A Journey with a Pilot*, reprint edition, 2016, 17.

278 "comfortable, well-fed" Anne Morrow Lindbergh, "Airliner to Europe," *Harper's Magazine*, September 1948, https://harpers.org/archive/1948/09/airliner-to-europe/.

279 "[t]here are no distant places" Micheline Maynard, "Prefer to Sit by the Window, Aisle or ATM?" *The Lede*, 1214616788, https://thelede.blogs.nytimes.com/2008/06 /27/prefer-to-sit-by-the-window-aisle-or-atm/.

279 "the right to go" Jenifer van Vleck, *Empire of the Air: Aviation and the American Ascendancy*, 2013, 90.

279 "the splendor of the world" John Rennie Short, *Globalization, Modernity and the City*, 2013, 142.

279 "[a]s the speed of aerial transit" Dave English, "Great Aviation Quotes: Predictions of the Future," http://aviationquotations.com//predictions.html, accessed May 1, 2017.

279 "The 'stratosphere' planes" Anne Morrow Lindbergh, *Listen! The Wind*, 1938, ix.

280 "We're not selling transportation" Emily Badger, "Why Even the Hyperloop Probably Wouldn't Change Your Commute Time," *New York Times*, August 10, 2017, https://www.nytimes.com/2017/08/10/upshot/why-even-the-hyperloop-probably -wouldnt-change-your-commute-time.html.

280 "The compass and the sextant" Nāʻālehu Anthony, presentation at "The *Hōkūleʻa*: Indigenous Resurgence from Hawaiʻi to Mannahatta," New York University, March 31, 2016.

281 "The GPS receiver's" Claudio Aporta and Eric Higgs, "Satellite Culture: Global Positioning Systems, Inuit Wayfinding, and the Need for a New Account of Technology," *Current Anthropology* 46, no. 5 (2005): 729–53, doi.org/10.1086/432651.

282 "I told him that" Ibid.

282 "the device paradigm" Albert Borgmann, *Technology and the Character of Contemporary Life: A Philosophical Inquiry*, 2009, 5.

283 "the machinery makes" Eric Higgs, Andrew Light, and David Strong, *Technology and the Good Life?* 2010, 29.

283 "The combination of newer" Aporta and Higgs, "Satellite Culture: Global Positioning Systems, Inuit Wayfinding, and the Need for a New Account of Technology," 729–53.

284 "Transport of the mails" Antoine de Saint-Exupéry, *Wind, Sand and Stars*, trans. Lewis Galantiere, 2002, 44.

284 "It is not with metal" Ibid., 43.

285 "If I can go a hundred miles" Joe Coughlin, presentation at "Planning Ideas That Matter, Faculty Debate: Part 3," MIT Department of Urban Studies and Planning, October 12, 2017, https://dusp.mit.edu/event/planning-ideas-matter-faculty-debate-part-3.

286 "dissembles its own" Ibid.

288 "will eventually find you" Elon Musk, tweet, @elonmusk, October 3, 2016, https://twitter.com/elonmusk/status/789022017311735808?lang=en.

288 "what is a sparrow" lostTesla, tweet, @*LostTesla* (blog), October 1, 2017, https://twitter.com/LostTesla/status/923979654704369664.

Epilogue

291 "[a]nalyzing landscape empowers" John R. Stilgoe, *What Is Landscape?* 2015, 49.

291 "[b]eing lost, even being" Ibid., xi.

292 "Men and boys built gliders" Ibid., 24.

293 "right to roam" David Derbyshire, "How Children Lost the Right to Roam in Four Generations," *Mail Online*, June 15, 2007, http://www.dailymail.co.uk/news/article-462091/How-children-lost-right-roam-generations.html.

293 "If we were on our" Helen E. Woolley and Elizabeth Griffin, "Decreasing Experiences of Home Range, Outdoor Spaces, Activities and Companions: Changes across Three Generations in Sheffield in North England," *Children's Geographies* 13, no. 6 (November 2, 2015): 677–91, doi.org/10.1080/14733285.2014.952186.

294 "Autonomy is a key" Ibid.

294 "there is no concept" Jean Piaget, *The Construction of Reality in the Child*, 2013, 98.

295 "genius of childhood" Edith Cobb, "The Ecology of Imagination in Childhood," *Daedalus* 88, no. 3 (1959): 537–48.

295 "This plasticity of response" Ibid.

295 "the spirit of place" Ibid.

296 "very important kind" James J. Gibson, *The Ecological Approach to Visual Perception: Classic Edition*, first edition, 2014, 229.

296 "for man, geographical reality" Edward Relph, *Place and Placelessness*, 1976, 11.

296 "Before any choice" Janet Donohoe, *Remembering Places: A Phenomenological Study of the Relationship between Memory and Place*, 2014, 12.

297 "Of course, peoples of the desert" Yi-Fu Tuan, *Topophilia: A Study of Environmental Perception, Attitudes, and Values*, reprint edition, 1990, xii.

298 "a whole system of predispositions" Sarah Gatson, "Habitus," *International Encyclopedia of the Social Sciences*, n.d., http://www.encyclopedia.com.

299 "think[ing] purely of the external world" Harold Gatty, *Finding Your Way Without Map or Compass*, reprint edition, 1999, 9.

299 "ask always the names" Stilgoe, *What Is Landscape?* 219.

299 "focal things" Eric Higgs, Andrew Light, and David Strong, *Technology and the Good Life?* 2010, 31.

300 "border-land of knowledge" Anna Botsford Comstock, *Handbook of Nature Study*, first edition, 1986, 4.

301 "perception and regard" Ibid., 1.

302 "In the context of massive" Guy Aitchison, "Do We All Have a Right to Cross Borders?" *The Conversation*, December 19, 2016, http://theconversation.com/do-we-all-have-a-right-to-cross-borders-69835.

302 "to be rooted is" Simone Weil, *The Need for Roots: Prelude to a Declaration of Duties Towards Mankind*, 2001, 43.

302 "certain particular treasures" Ibid.

303 "an overwhelming, unexchangeable" Relph, *Place and Placelessness*.

303 "persistent thinking of home" W. H. McCann, "Nostalgia: A Review of the Litera-
      ture," *Psychological Bulletin* 38, no. 3 (March 1, 1941): 165–82.

303 "the chirp of crickets" Ibid.

304 "mother centering tendency" Ibid.

304 "This century has" Robyn Davidson, *Desert Place*, first edition, 1996. 5.

SELECTED BIBLIOGRAPHY

Abazov, Rafis. "Globalization of Migration: What the Modern World Can Learn from Nomadic Cultures." UN Chronicle, September 2013. https://unchronicle.un .org/article/globalization-migration-what-modern-world-can-learn-nomadic -cultures.

Aitchison, Guy. "Do We All Have a Right to Cross Borders?" *The Conversation*, December 19, 2016. http://theconversation.com/do-we-all-have-a-right-to-cross-borders -69835.

Alerstam, Thomas. "Conflicting Evidence about Long-Distance Animal Navigation." *Science* 313, no. 5788 (August 11, 2006): 791–94. https://doi.org/10.1126/science.1129048.

Altman, Irwin, and Setha M. Low. *Place Attachment*. Berlin: Springer Science & Business Media, 2012.

Anthony, Nāʻālehu. Presentation at "The *Hōkūleʻa*: Indigenous Resurgence from Hawaiʻi to Mannahatta." New York University, March 31, 2016.

Aporta, Claudio. "Inuit Orienting: Traveling along Familiar Horizons." *Sensory Studies*, n.d. http://www.sensorystudies.org/inuit-orienting-traveling-along-familiar-horizons/.

———. "Old Routes, New Trails: Contemporary Inuit Travel and Orienting in Igloolik, Nunavut." Thesis, University of Alberta, 2003.

Aporta, Claudio, and Eric Higgs. "Satellite Culture: Global Positioning Systems, Inuit Wayfinding, and the Need for a New Account of Technology." *Current Anthropology* 46, no. 5 (2005): 729–53. https://doi.org/10.1086/432651.

Aporta, Claudio, and Nunavut Research Institute. *Anijaarniq: Introducing Inuit Landskills and Wayfinding*. Iqaluit: Nunavut Research Institute, 2006.

"Apprentice Maui Navigator Takes Helm of *Hikianalia* in Voyage to Tahiti." *Maui Now*, March 22, 2017. http://mauinow.com/2017/03/22/apprentice-maui-navigator-takes -helm-of-hikianalia-in-voyage-to-tahiti/.

Ardila, A. "Historical Evolution of Spatial Abilities." *Behavioural Neurology* 6, no. 2 (1993): 83–87. https://doi.org/10.3233/BEN-1993-6203.

Badger, Emily. "The Surprisingly Complex Art of Urban Wayfinding." *CityLab*, 2012. http://www.theatlanticcities.com/design/2012/01/surprisingly-complex-art-way finding/1088/.

———. "Why Even the Hyperloop Probably Wouldn't Change Your Commute Time." *New York Times*, August 10, 2017, sec. The Upshot. https://www.nytimes.com/2017

/08/10/upshot/why-even-the-hyperloop-probably-wouldnt-change-your-commute
-time.html.

Baird, Benjamin, Jonathan Smallwood, and Jonathan W. Schooler. "Back to the Future: Autobiographical Planning and the Functionality of Mind-Wandering." *Consciousness and Cognition, From Dreams to Psychosis* 20, no. 4 (December 1, 2011): 1604–11. https://doi.org/10.1016/j.concog.2011.08.007.

Baker, R. Robin. *Human Navigation and the Sixth Sense.* First edition. New York: Simon & Schuster, 1981.

Bar, Moshe, ed. *Predictions in the Brain: Using Our Past to Generate a Future.* Revised edition. New York: Oxford University Press, 2011.

Barger, Nicole, Kari L. Hanson, Kate Teffer, Natalie M. Schenker-Ahmed, and Katerina Semendeferi. "Evidence for Evolutionary Specialization in Human Limbic Structures." *Frontiers in Human Neuroscience* 8 (May 20, 2014). https://doi.org/10.3389 /fnhum.2014.00277.

Barton, Phillip Lionel. "Maori Cartography and the European Encounter." In *The History of Cartography: Cartography in the Traditional African, American, Arctic, Australian, and Pacific Societies,* edited by David Woodward and G. Malcolm Lewis, vol. 2, book 3. 493–532. First edition. Chicago: University of Chicago Press, 1998.

Basso, Keith H. *Wisdom Sits in Places: Landscape and Language among the Western Apache.* Albuquerque: University of New Mexico Press, 1996.

Beaglehole, J. C. *The Life of Captain James Cook.* Palo Alto, CA: Stanford University Press, 1992.

Bennardo, Giovanni. "Linguistic Relativity and Spatial Language." *Linguistic Anthropology* (2009): 137.

Bennett, John, and Susan Rowley. *Uqalurait: An Oral History of Nunavut.* Montreal: McGill-Queen's University Press, 2004.

Berg, Mary, and Elliott A. Medrich. "Children in Four Neighborhoods." *Environment and Behavior* 12, no. 3 (1980): 320–48. http://journals.sagepub.com/doi/abs/10.1177 /0013916580123003.

Berman, Bradley. "Whoever Owns the Maps Owns the Future of Self-Driving Cars." *Popular Mechanics,* July 1, 2016. http://www.popularmechanics.com/cars/a21609 /here-maps-future-of-self-driving-cars/.

Berthold, Peter. *Bird Migration: A General Survey.* Oxford: Oxford University Press, 2001.

Berwick, Robert C., Angela D. Friederici, Noam Chomsky, and Johan J. Bolhuis. "Evolution, Brain, and the Nature of Language." *Trends in Cognitive Sciences* 17, no. 2 (February 1, 2013): 89–98. https://doi.org/10.1016/j.tics.2012.12.002.

Berwick, Robert C., and Noam Chomsky. *Why Only Us: Language and Evolution.* Cambridge, MA: MIT Press, 2016.

Bickerton, Derek. *More Than Nature Needs: Language, Mind, and Evolution.* Cambridge, MA: Harvard University Press, 2014.

Blades, Mark. "Children's Ability to Learn about the Environment from Direct Experience and from Spatial Representations." *Children's Environments Quarterly* 6, nos. 2/3 (1989): 4–14.

Boggs, James P. "The Culture Concept as Theory, in Context." *Current Anthropology* 45, no. 2 (2004). http://www.journals.uchicago.edu/doi/abs/10.1086/381048.

Bohbot, Véronique D. "All Roads Lead to Rome, Even in African Savannah Elephants— or Do They?" *Proceedings of the Royal Society B,* 2015. http://www.bic.mni.mcgill .ca/users/vero/PAPERS/Bohbot2015.pdf.

Bohbot, Véronique D., Jason Lerch, Brook Thorndycraft, Giuseppe Iaria, and Alex P. Zijdenbos. "Gray Matter Differences Correlate with Spontaneous Strategies in a Human Virtual Navigation Task." *Journal of Neuroscience* 27, no. 38 (September 19, 2007): 10078–83. https://doi.org/10.1523/JNEUROSCI.1763-07.2007.

Borgmann, Albert. *Technology and the Character of Contemporary Life: A Philosophical Inquiry.* Chicago: University of Chicago Press, 2009.

———. "Technology as a Cultural Force: For Alena and Griffin." *Canadian Journal of Sociology* 31, no. 3 (September 6, 2006): 351–60. https://doi.org/10.1353/cjs.2006.0050.

Boroditsky, Lera. "Does Language Shape Thought? Mandarin and English Speakers' Conceptions of Time." *Cognitive Psychology* 43, no. 1 (August 1, 2001): 1–22. https://doi.org/10.1006/cogp.2001.0748.

———. "Metaphoric Structuring: Understanding Time through Spatial Metaphors." *Cognition* 75, no. 1 (April 14, 2000): 1–28. https://doi.org/10.1016/S0010-0277(99)00073-6.

Boudette, Neal E. "Building a Road Map for the Self-Driving Car." *New York Times*, March 2, 2017, sec. Automobiles. https://www.nytimes.com/2017/03/02/automobiles/wheels/self-driving-cars-gps-maps.html.

Bower, Gordon H. "Analysis of a Mnemonic Device: Modern Psychology Uncovers the Powerful Components of an Ancient System for Improving Memory." *American Scientist* 58, no. 5 (1970): 496–510.

Brachmayer, Scott. *Kajutaijuq.* Film short, 2015. http://www.imdb.com/title/tt3826696/.

Bradley, C. E. "Traveling with Fred George: The Changing Ways of Yup'ik Star Navigation in Akiachak Western Alaska." In *The Earth Is Faster Now: Indigenous Observations of Arctic Environmental Change: Frontiers in Polar Social Science*, edited by Igor Krupnik and Dyanna Jolly, 240–265. Fairbanks: Arctic Research Consortium of the United States, 2002.

Briggs, Jean L. *Inuit Morality Play: The Emotional Education of a Three-Year-Old.* New Haven, CT: Yale University Press, 1998.

Brown, Frank A., J. Woodland Hastings, and John D. Palmer. *The Biological Clock: Two Views.* Cambridge: Academic Press, 2014.

Brownell, Ginanne. "Looking Forward, Fiji Turns to Its Canoeing Past." *New York Times*, February 3, 2012, sec. Global Business. https://www.nytimes.com/2012/02/04/business/global/looking-forward-fiji-turns-to-its-canoeing-past.html.

Buchanan, Brett. *Onto-Ethologies: The Animal Environments of Uexküll, Heidegger, Merleau-Ponty, and Deleuze.* Albany: State University of New York Press, 2009.

Bullens, Jessie, Kinga Iglói, Alain Berthoz, Albert Postma, and Laure Rondi-Reig. "Developmental Time Course of the Acquisition of Sequential Egocentric and Allocentric Navigation Strategies." *Journal of Experimental Child Psychology* 107, no. 3 (November 1, 2010): 337–50. https://doi.org/10.1016/j.jecp.2010.05.010.

Burda, Hynek, Sabine Begall, Jaroslav Červený, Julia Neef, and Pavel Němec. "Extremely Low-Frequency Electromagnetic Fields Disrupt Magnetic Alignment of Ruminants." *Proceedings of the National Academy of Sciences* 106, no. 14 (April 7, 2009): 5708–13. https://doi.org/10.1073/pnas.0811194106.

Burgess, Neil. "Spatial Memory: How Egocentric and Allocentric Combine." *Trends in Cognitive Sciences* 10, no. 12 (December 1, 2006): 551–57. https://doi.org/10.1016/j.tics.2006.10.005.

Burgess, Neil, Eleanor A. Maguire, and John O'Keefe. "The Human Hippocampus and Spatial and Episodic Memory." *Neuron* 35, no. 4 (2002): 625–41.

Burgess, Neil, Hugo J. Spiers, and Eleni Paleologou. "Orientational Manoeuvres in the Dark: Dissociating Allocentric and Egocentric Influences on Spatial Memory." *Cognition* 94, no. 2 (December 2004): 149–66. https://doi.org/10.1016/j.cognition.2004.01.001.

Burke, Ariane. "Spatial Abilities, Cognition and the Pattern of Neanderthal and Modern Human Dispersals." In "The Neanderthal Home: Spatial and Social Behaviours." Special issue, *Quaternary International* 247, no. Supplement C (January 9, 2012): 230–35. https://doi.org/10.1016/j.quaint.2010.10.029.

Burke, Ariane, Anne Kandler, and David Good. "Women Who Know Their Place." *Human Nature* 23, no. 2 (June 1, 2012): 133–48. https://doi.org/10.1007/s12110-012-9140-1.

Burnett, G. E., and Kate Lee. "The Effect of Vehicle Navigation Systems on the Formation of Cognitive Maps." In *Traffic and Transport Psychology*, edited by G. Under-
  wood, 407–17. Amsterdam: Elsevier Science, 2005.
Buss, Irven O. "Bird Detection by Radar." *The Auk* 63, no. 3 (1946): 315–18. https://doi
  .org/10.2307/4080116.
Cairns, Hugh. *Dark Sparklers: Yidumduma's Wardaman Aboriginal Astronomy Northern Australia 2003*. Merimbula, NSW, Australia: H. C. Cairns, 2003.
Callaghan, Bridget L., Stella Li, and Rick Richardson. "The Elusive Engram: What Can
  Infantile Amnesia Tell Us about Memory?" *Trends in Neurosciences* 37, no. 1 (Janu-
  ary 1, 2014): 47–53. https://doi.org/10.1016/j.tins.2013.10.007.
Cane, Scott. *First Footprints: The Epic Story of the First Australians*. Main edition. Syd-
  ney: Allen & Unwin, 2014.
Caruana, Wally. *Aboriginal Art*. Third edition. London: Thames & Hudson, 2013.
Casasanto, Daniel. "Space for Thinking." In *Language, Cognition and Space*. London:
  Equinox, n.d.
Chaddock, Laura, et al. "A Neuroimaging Investigation of the Association between Aer-
  obic Fitness, Hippocampal Volume, and Memory Performance in Preadolescent
  Children." *Brain Research* 1358 (October 28, 2010): 172–83. https://doi.org/10.1016
  /j.brainres.2010.08.049.
Chadwick, Martin J., and Hugo J Spiers. "A Local Anchor for the Brain's Compass." *Nature
  Neuroscience* 17, no. 11 (November 2014). https://www.nature.com/articles/nn.3841.
Chatty, Dawn. *Nomadic Societies in the Middle East and North Africa: Entering the
  21st Century*. Leiden: Brill, 2006.
Chawla, Louise. "Ecstatic Places." In *The People, Place, and Space Reader*, edited by Jen
  Jack Gieseking, William Mangold, and Cindi Katz. London: Routledge, 2014.
Chen, Chuansheng, Michael Burton, Ellen Greenberger, and Julia Dmitrieva. "Popula-
  tion Migration and the Variation of Dopamine D4 Receptor (DRD4) Allele Frequen-
  cies Around the Globe." *Evolution and Human Behavior* 20, no. 5 (1999): 309–24.
Clynes, Tom. "How to Raise a Genius: Lessons from a 45-Year Study of Super-Smart
  Children." *Nature News* 537, no. 7619 (September 8, 2016): 152. https://doi.org/10
  .1038/537152a.
Cobb, Edith. "The Ecology of Imagination in Childhood." *Daedalus* 88, no. 3 (1959): 537–48.
Cobb, Matthew. "Are We Ready for Quantum Biology?" *New Scientist*, November 12,
  2014. https://www.newscientist.com/article/mg22429950-700-are-we-ready-for-quan
  tum-biology/.
Cohen, Neal J., and Howard Eichenbaum. *Memory, Amnesia, and the Hippocampal Sys-
  tem*. New edition. A Bradford Book. Cambridge, MA: MIT Press, 1995.
Collett, Thomas S., and Paul Graham. "Animal Navigation: Path Integration, Visual
  Landmarks and Cognitive Maps." *Current Biology* 14, no. 12 (June 22, 2004): R475–
  77. https://doi.org/10.1016/j.cub.2004.06.013.
Comstock, Anna Botsford. *Handbook of Nature Study*. First edition. Ithaca, NY: Com-
  stock/Cornell University Press, 1986.
Convit, A., M. J. De Leon, C. Tarshish, S. De Santi, W. Tsui, H. Rusinek, and A. George.
  "Specific Hippocampal Volume Reductions in Individuals at Risk for Alzheimer's
  Disease." *Neurobiology of Aging* 18, no. 2 (March 1, 1997): 131–38. https://doi.org/10
  .1016/S0197-4580(97)00001-8.
Corballis, Michael C. "Mental Time Travel: A Case for Evolutionary Continuity." *Trends
  in Cognitive Sciences* 17, no. 1 (January 2013): 5–6. https://doi.org/10.1016/j.tics.2012
  .10.009.
Coughlin, Joe. "Planning Ideas That Matter, Faculty Debate: Part 3." MIT Department
  of Urban Studies and Planning, October 12, 2017. https://dusp.mit.edu/event/planning
  -ideas-matter-faculty-debate-part-3.
Coutrot, Antoine, et al. "Global Determinants of Navigation Ability." *bioRxiv* (September 18, 2017): 188870. https://doi.org/10.1101/188870.

Crawford, Matthew B. *The World beyond Your Head: On Becoming an Individual in an Age of Distraction*. Reprint edition. New York: Farrar, Straus and Giroux, 2016.

Cristea, Anca, David Hummels, Laura Puzzello, and Misak Avetisyan. "Trade and the Greenhouse Gas Emissions from International Freight Transport." *Journal of Environmental Economics and Management* 65, no. 1 (January 1, 2013): 153–73. https://doi.org/10.1016/j.jeem.2012.06.002.

Curry, Andrew. "Men Are Better at Maps until Women Take This Course." *Nautilus*, January 28, 2016. http://nautil.us/issue/32/space/men-are-better-at-maps-until-women-take-this-course.

Cyranoski, David. "Discovery of Long-Sought Biological Compass Claimed." *Nature* 527, no. 7578 (November 16, 2015): 283–84. https://doi.org/10.1038/527283a.

Dardel, Éric. *L'homme et la terre: Nature de la réalité géographique*. Paris: Editions du CTHS, 1990.

Darwin, Charles. *Charles Darwin's Shorter Publications, 1829–1883*. Cambridge: Cambridge University Press, 2009.

———. "Origin of Certain Instincts." *Nature* 7, no. 179 (April 3, 1873): 007417a0. https://doi.org/10.1038/007417a0.

Davidson, Robyn. *Desert Places*. First edition. New York: Viking, 1996.

———. *Quarterly Essay 24: No Fixed Address: Nomads and the Fate of the Planet*. Melbourne, Black Inc, 2006.

Dearden, Lizzie. "Syrian Refugee Tells How He Survived Boat Sinking in Waters Where Aylan Kurdi Drowned." *The Independent*, September 3, 2015. https://www.independent.co.uk/news/world/europe/syrian-refugee-tells-how-he-survived-boat-sinking-in-waters-where-aylan-kurdi-drowned-10484607.html.

De Leon, M. J., et al. "Frequency of Hippocampal Formation Atrophy in Normal Aging and Alzheimer's Disease." *Neurobiology of Aging* 18, no. 1 (January 1, 1997): 1–11. https://doi.org/10.1016/S0197-4580(96)00213-8.

Delmore, Kira, et al. "Genomic Analysis of a Migratory Divide Reveals Candidate Genes for Migration and Implicates Selective Sweeps in Generating Islands of Differentiation." *Molecular Ecology* 24, no. 8 (April 3, 2015). http://onlinelibrary.wiley.com/doi/10.1111/mec.13150/full.

Derbyshire, David. "How Children Lost the Right to Roam in Four Generations." *Mail Online*, June 15, 2007. http://www.dailymail.co.uk/news/article-462091/How-children-lost-right-roam-generations.html.

Devlin, Hannah. "Google Creates AI Program That Uses Reasoning to Navigate the London Tube." *The Guardian*, October 12, 2016, sec. Technology. https://www.theguardian.com/technology/2016/oct/12/google-creates-ai-program-that-uses-reasoning-to-navigate-the-london-tube.

Diaz, Vicente. "No Island Is an Island." In *Native Studies Keywords*, edited by Stephanie Nohelani Teves, Andrea Smith, and Michelle Raheja, 90–107. Critical Issues in Indigenous Studies. Tucson: University of Arizona Press, 2015.

———. Presentation at "The *Hōkūle'a*: Indigenous Resurgence from Hawai'i to Mannahatta." New York University, March 31, 2016.

———. "Lost in Translation and Found in Constipation: Unstopping the Flow of Intangible Cultural Heritage with the Embodied Tangibilities of Traditional Carolinian Seafaring Culture." International Symposium on Negotiating Intangible Cultural Heritage, National Ethnology Museum, Osaka, Japan, 2017.

Dickinson, Anthony, and Nicola S. Clayton. "Episodic-Like Memory during Cache Recovery by Scrub Jays." *Nature* 395, no. 6699 (September 17, 1998): 272. https://doi.org/10.1038/26216.

Dingle, Hugh. "Animal Migration: Is There a Common Migratory Syndrome?" *Journal of Ornithology* 147, no. 2 (April 1, 2006): 212–20. https://doi.org/10.1007/s10336-005-0052-2.

———. *Migration: The Biology of Life on the Move*. First edition. New York: Oxford University Press, 1996.

Di Piazza, Anne. "A Reconstruction of a Tahitian Star Compass Based on Tupaia's 'Chart for the Society Islands with Otaheite in the Center.'" *Journal of the Polynesian Society* 119, no. 4 (2010): 377–92.

Dodge, Richard Irving, and General William Tecumseh Sherman. *Our Wild Indians: Thirty-Three Years' Personal Experience among the Red Men of the Great West—A Popular Account of Their Social Life, Religion, Habits, Traits, Customs, Exploits, Etc.* Reprint edition. Williamstown, MA: Corner House Pub, 1978.

Donohoe, Janet. *Remembering Places: A Phenomenological Study of the Relationship between Memory and Place*. Lanham, MD: Lexington Books, 2014.

Dorais, Louis-Jacques. *The Language of the Inuit: Syntax, Semantics, and Society in the Arctic*. Montreal: McGill-Queen's University Press, 2010.

Dresler, Martin, et al. "Mnemonic Training Reshapes Brain Networks to Support Superior Memory." *Neuron* 93, no. 5 (March 8, 2017): 1227–35.e6. https://doi.org/10.1016/j.neuron.2017.02.003.

Druett, Joan. *Tupaia: Captain Cook's Polynesian Navigator*. Santa Barbara, CA: Praeger, 2010.

Dyer, Fred C., and James L. Could. "Honey Bee Navigation: The Honey Bee's Ability to Find Its Way Depends on a Hierarchy of Sophisticated Orientation Mechanisms." *American Scientist* 71, no. 6 (1983): 587–97.

Eber, Dorothy Harley. *Encounters on the Passage: Inuit Meet the Explorers*. Second edition. Toronto: University of Toronto Press, Scholarly Publishing Division, 2008.

Edwards, Tim. "How I Got Home." *Up Here Magazine*, February 10, 2016. https://uphere.ca/articles/how-i-got-home.

Eichenbaum, Howard. "Hippocampus: Mapping or Memory?" *Current Biology* 10, no. 21 (November 1, 2000): R785–87. https://doi.org/10.1016/S0960-9822(00)00763-6.

Eichenbaum, Howard, and Neal J. Cohen. "Can We Reconcile the Declarative Memory and Spatial Navigation Views on Hippocampal Function?" *Neuron* 83, no. 4 (August 20, 2014): 764–70. https://doi.org/10.1016/j.neuron.2014.07.032.

Eichenbaum, Howard, Paul Dudchenko, Emma Wood, Matthew Shapiro, and Heikki Tanila. "The Hippocampus, Memory, and Place Cells." *Neuron* 23, no. 2 (June 1999). http://www.cell.com/fulltext/S0896-6273(00)80773-4.

Evans, Vyvyan, and Paul Chilton, eds. *Language, Cognition, and Space: The State of the Art and New Directions*. London: Equinox, 2010.

Fara, Patricia. "An Atttractive Therapy: Animal Magnetism in Eighteenth-Century England." *History of Science* 33, no. 2 (June 1, 1995): 127–77. https://doi.org/10.1177/007327539503300201.

Fent, Karl, and Rudiger Wehner. "Ocelli: A Celestial Compass in the Desert Ant Cataglyphis." *Science* 228, no. 4696 (April 12, 1985): 192–94. https://doi.org/10.1126/science.228.4696.192.

Finney, Ben. "Nautical Cartography and Traditional Navigation in Oceania." In *The History of Cartography: Cartography in the Traditional African, American, Arctic, Australian, and Pacific Societies*, edited by David Woodward and G. Malcolm Lewis, vol. 2, book 3, 443–492. Chicago: University of Chicago Press, 1998.

———. *Voyage of Rediscovery: A Cultural Odyssey through Polynesia*. Berkeley: University of California Press, 1994.

Fladmark, J. M., and Thor Heyerdahl. *Heritage and Identity: Shaping the Nations of the North*. London: Routledge, 2015.

Fledler, Nadine. "The Hidden Gringo." *Reed Magazine*, August 2000. http://www.reed.edu/reed_magazine/aug2000/a_gringo/3.html.

Fleur, Nicholas St. "How Ancient Humans Reached Remote South Pacific Islands." *New York Times*, November 1, 2016. http://www.nytimes.com/2016/11/02/science/south-pacific-islands-migration.html.

Ford, James D., Barry Smit, Johanna Wandel, and John MacDonald. "Vulnerability to Climate Change in Igloolik, Nunavut: What We Can Learn from the Past and Present." *Polar Record* 42, no. 2 (April 2006): 127–38. https://doi.org/10.1017/S00322474060 05122.

Fortescue, Michael. *Eskimo Orientation Systems*. Copenhagen: Museum Tusculanum Press, 1988.

Frake, Charles O. "Cognitive Maps of Time and Tide among Medieval Seafarers." *Man* 20, no. 2 (1985): 254–70. https://doi.org/10.2307/2802384.

Freud, Sigmund. *Freud on Women: A Reader*. New York: W. W. Norton, 1992.

Friedman, Uri. "A World of Walls." *The Atlantic*, May 19, 2016. https://www.theatlantic .com/international/archive/2016/05/donald-trump-wall-mexico/483156/.

Friesen, Max. "North America: Paleoeskimo and Inuit Archaeology," in *Encyclopedia of Global Human Migration*. Edited by Immanuel Ness. Hoboken, NJ: Blackwell Publishing, 2013. https://www.academia.edu/5314092/North_America_Paleoeskimo _and_Inuit_archaeology_Encyclopedia_of_Global_Human_Migration_.

Gammage, Bill. *The Biggest Estate on Earth: How Aborigines Made Australia*. Reprint edition. Crows Nest, NSW: Allen & Unwin, 2013.

Gatson, Sarah. "Habitus." *International Encyclopedia of the Social Sciences*, n.d. http:// www.encyclopedia.com.

Gatty, Harold. *Finding Your Way Without Map or Compass*. Reprint edition. Mineola, NY: Dover Publications, 1999.

Gatty, Harold, and J. H. Doolittle. *Nature Is Your Guide: How to Find Your Way on Land and Sea by Observing Nature*. New York: Penguin Books, 1979.

Gell, Alfred. "How to Read a Map: Remarks on the Practical Logic of Navigation." *Man* 20, no. 2 (June 1985): 271–86. https://doi.org/10.2307/2802385.

———. "Vogel's Net: Traps as Artworks and Artworks as Traps." *Journal of Material Culture* (1996): 15–38.

Gentner, Dedre, Mutsumi Imai, and Lera Boroditsky. "As Time Goes By: Evidence for Two Systems in Processing Space → Time Metaphors." *Language and Cognitive Processes* 17, no. 5 (October 1, 2002): 537–65. https://doi.org/10.1080/01690960143000317.

Genz, Joseph. "Navigating the Revival of Voyaging in the Marshall Islands: Predicaments of Preservation and Possibilities of Collaboration." *Contemporary Pacific* 23, no. 1 (March 26, 2011): 1–34. https://doi.org/10.1353/cp.2011.0017.

Genz, Joseph, Jerome Aucan, Mark Merrifield, Ben Finney, Korent Joel, and Alson Kelen. "Wave Navigation in the Marshall Islands: Comparing Indigenous and Western Scientific Knowledge of the Ocean." *Oceanography* 22, no. 2 (2009): 234–45. http:// agris.fao.org/agris-search/search.do?recordID=DJ2012092628.

Gibson, E. J. "Perceptual Learning." *Annual Review of Psychology* 14, no. 1 (1963): 29– 56. https://doi.org/10.1146/annurev.ps.14.020163.000333.

Gibson, J. J., and E. J. Gibson. "Perceptual Learning: Differentiation or Enrichment?" *Psychological Review* 62, no. 1 (January 1955): 32–41.

Gibson, James J. *The Ecological Approach to Visual Perception: Classic Edition*. First edition. New York: Psychology Press, 2014.

———. *The Senses Considered as Perceptual Systems*. Revised edition. Westport, CT: Praeger, 1983.

Gill, Victoria. "Great Monarch Butterfly Migration Mystery Solved." BBC News, 2016. http://www.bbc.com/news/science-environment-36046746.

Ginzburg, Carlo. *Clues, Myths, and the Historical Method*. Translated by John Tedeschi and Anne C. Tedeschi. Reprint edition. Baltimore: Johns Hopkins University Press, 2013.

Gladwin, Thomas. *East Is a Big Bird: Navigation and Logic on Puluwat Atoll*. Cambridge, MA: Harvard University Press, 1995.

Golledge, Reginald G. *Spatial Behavior: A Geographic Perspective*. New York: Guilford Press, 1997.

———, ed. *Wayfinding Behavior: Cognitive Mapping and Other Spatial Processes.* First edition. Baltimore: Johns Hopkins University Press, 1998.

Golledge, Reginald G., Nathan Gale, James W. Pellegrino, and Sally Doherty. "Spatial Knowledge Acquisition by Children: Route Learning and Relational Distances." *Annals of the Association of American Geographers* 82, no. 2 (June 1, 1992): 223–44. https://doi.org/10.1111/j.1467-8306.1992.tb01906.x.

Goodman, Russell. "William James." In *The Stanford Encyclopedia of Philosophy*, edited by Edward N. Zalta. Metaphysics Research Lab, Stanford University, Winter 2017. https://plato.stanford.edu/archives/win2017/entries/james/.

Goodwin, James C. "A-Mazing Research." *American Psychological Association* 43, no. 2 (February 2012). http://www.apa.org/monitor/2012/02/research.aspx.

Gooley, Tristan. *The Lost Art of Reading Nature's Signs: Use Outdoor Clues to Find Your Way, Predict the Weather, Locate Water, Track Animals—and Other Forgotten Skills.* Reprint edition. New York: The Experiment, 2015.

———. *The Natural Navigator: The Rediscovered Art of Letting Nature Be Your Guide.* Reprint edition. New York: The Experiment, 2012.

———. "The Navigator That Time Lost." The Natural Navigator, 2009. https://www.naturalnavigator.com/the-library/the-navigator-that-time-lost.

Gould, James L. "Animal Navigation: Memories of Home." *Current Biology* 25, no. 3 (February 2, 2015): R104–6. https://doi.org/10.1016/j.cub.2014.12.024.

Gould, James L., and Carol Grant Gould. *Nature's Compass: The Mystery of Animal Navigation.* Princeton, NJ: Princeton University Press, 2012.

Grabar, Henry. "Smartphones and the Uncertain Future of 'Spatial Thinking.'" *CityLab*, 2014. http://www.citylab.com/tech/2014/09/smartphones-and-the-uncertain-future-of-spatial-thinking/379796/.

Griffiths, Daniel, Anthony Dickinson, and Nicola Clayton. "Episodic Memory: What Can Animals Remember about Their Past?" *Trends in Cognitive Sciences* 3, no. 2 (February 1, 1999): 74–80. https://doi.org/10.1016/S1364-6613(98)01272-8.

Guiducci, Dario, and Ariane Burke. "Reading the Landscape: Legible Environments and Hominin Dispersals." *Evolutionary Anthropology* 25, no. 3 (May 6, 2016): 133–41. https://doi.org/10.1002/evan.21484.

Guilford, Tim, et al. "Migratory Navigation in Birds: New Opportunities in an Era of Fast-Developing Tracking Technology." *Journal of Experimental Biology* 214, no. 22 (November 15, 2011): 3705–12. https://doi.org/10.1242/jeb.051292.

Gumperz, John J., and Stephen C. Levinson. *Rethinking Linguistic Relativity.* Cambridge: Cambridge University Press, 1996.

Gupta, Akhil, and James Ferguson. "Beyond 'Culture': Space, Identity, and the Politics of Difference." *Cultural Anthropology* 7, no. 1 (1992): 6–23.

Hallendy, Norman. *Inuksuit: Silent Messengers of the Arctic.* First trade paper edition. Vancouver, BC: Douglas & McIntyre, 2001.

———. *Tukiliit: The Stone People Who Live in the Wind; An Introduction to Inuksuit and Other Stone Figures of the North.* Vancouver, BC: Douglas & McIntyre, 2009. https://docslide.com.br/documents/tukiliit-the-stone-people-who-live-in-the-wind-an-introduction-to-inuksuit.html.

Harley, J. B., and David Woodward, eds. *The History of Cartography: Cartography in the Traditional Islamic and South Asian Societies.* Vol. 2, book 1. First edition. Chicago: University of Chicago Press, 1992.

Harney, William Edward. *Life among the Aborigines.* London: R. Hale, 1957.

Harper, Kenn. "Wooden Maps." *Nunatsiaq Online*, April 11, 2014. http://www.nunatsiaq.com/stories/article/65674taissumani_april_11/.

Haviland, John B. "Guugu Yimithirr Cardinal Directions." *Ethos* 26, no. 1 (March 1, 1998): 25–47. https://doi.org/10.1525/eth.1998.26.1.25.

Heft, Harry. "The Ecological Approach to Navigation: A Gibsonian Perspective." In

The Construction of Cognitive Maps, edited by J. Portugali, 105–32. GeoJournal Library. Dordrecht: Springer, 1996. https://doi.org/10.1007/978-0-585-33485-1_6.

———. *Ecological Psychology in Context: James Gibson, Roger Barker, and the Legacy of William James*. New York: Psychology Press, 2005.

———. "Way-Finding, Navigation and Environmental Cognition from a Naturalist's Stance." In *Handbook of Spatial Cognition*, edited by David Waller and Lynn Nadel. Washington, DC: APA Books, 2012.

Herbert, Jane, Julien Gross, and Harlene Hayne. "Crawling Is Associated with More Flexible Memory Retrieval by 9-Month-Old Infants." *Developmental Science* 10, no. 2 (March 1, 2007): 183–89. https://doi.org/10.1111/j.1467-7687.2007.00548.x.

Herculano-Houzel, Suzana, et al. "The Elephant Brain in Numbers." *Frontiers in Neuroanatomy* 8 (June 12, 2014). https://doi.org/10.3389/fnana.2014.00046.

Hercus, Luise, Jane Simpson, and Flavia Hodges. *The Land Is a Map: Placenames of Indigenous Origin in Australia*. Canberra: ANU Press, 2009. http://www.oapen.org/search?identifier=459353.

Hewitt, John. "Can a 'Quantum Compass' Help Birds Navigate via Magnetic Field?" *ExtremeTech*, March 2, 2015. http://www.extremetech.com/extreme/200051-can-a-biological-quantum-compass-help-birds-navigate-via-magnetic-field.

Hibar, D. P., et al. "Novel Genetic Loci Associated with Hippocampal Volume." *Nature Communications* 8 (January 18, 2017). http://dx.doi.org/10.1038/ncomms13624.

Hibar, Derrek P., et al. "Common Genetic Variants Influence Human Subcortical Brain Structures." *Nature* 520, no. 7546 (April 2015): 224–29. https://doi.org/10.1038/nature14101.

Higgs, Eric, Andrew Light, and David Strong. *Technology and the Good Life?* Chicago: University of Chicago Press, 2010.

Hill, Kenneth. "The Psychology of Lost." *Lost Person Behavior*. Ottawa: National SAR Secretariat, 1998, 1–16.

Hochmair, Hartwig H., and Klaus Luttich. "An Analysis of the Navigation Metaphor—and Why It Works for the World Wide Web." *Spatial Cognition & Computation* 6, no. 3 (September 1, 2006): 235–78. https://doi.org/10.1207/s15427633scc0603_3.

Holbrook, Jarita C. "Celestial Navigation and Technological Change on Moce Island." In *The Globalization of Knowledge in History*, edited by Jürgen Renn, 439–58. Berlin: epubli, 2012.

Holland, Elisabeth, et al. "Connecting the Dots: Policy Connections between Pacific Island Shipping and Global CO_2 and Pollutant Emission Reduction." *Carbon Management* 5, no. 1 (February 1, 2014): 93–105. https://doi.org/10.4155/cmt.13.78.

Horton, Travis W., et al. "Straight as an Arrow: Humpback Whales Swim Constant Course Tracks during Long-Distance Migration." *Biology Letters* (April 20, 2011): rsbl20110279. https://doi.org/10.1098/rsbl.2011.0279.

Hutchins, Edwin. *Cognition in the Wild*. Revised edition. A Bradford Book. Cambridge, MA: MIT Press, 1996.

Huth, John Edward. "Losing Our Way in the World." *New York Times*, July 20, 2013, sec. Opinion. http://www.nytimes.com/2013/07/21/opinion/sunday/losing-our-way-in-the-world.html.

———. *The Lost Art of Finding Our Way*. Reprint edition. Cambridge, MA: Belknap Press, 2015.

Ingold, Tim. "From Science to Art and Back Again: The Pendulum of an Anthropologist." *Anuac* 5, no. 1 (2016): 5–23.

———. *The Perception of the Environment: Essays on Livelihood, Dwelling and Skill*. First edition. London: Routledge, 2011.

———. "Up, Across and Along." *Place and Location: Studies in Environmental Aesthetics and Semiotics* 5 (2006): 21–36.

Ingold, Tim, Ana Letícia de Fiori, José Agnello Alves Dias de Andrade, Adriana Queiróz Testa, and Yuri Bassichetto Tambucci. "Wayfaring Thoughts: Life, Movement and

Anthropology." *Ponto Urbe: Revista Do Núcleo de Antropologia Urbana Da USP*, no. 11 (December 1, 2012). https://doi.org/10.4000/pontourbe.341.

"Introduction: Place Names in Nunarat." Inuit Heritage Trust: Place Names Program, n.d. http://ihti.ca/eng/place-names/pn-index.html?agree=0.

Inuit Place Names Project. "Where We Live and Travel: Named Places and Selected Routes." n.d. http://ihti.ca/eng/place-names/images/Map-WhereWeLiveTravel-1636px.jpg.

"Inuit Taxonomy." Canada's Polar Life: University of Guelph, n.d. http://www.arctic.uoguelph.ca/cpl/Traditional/class_frame.htm.

Ipeelee, Arnaitok. "The Old Ways of the Inuit." *Tumivut*, 1995.

Ishikawa, Toru, Hiromichi Fujiwara, Osamu Imai, and Atsuyuki Okabe. "Wayfinding with a GPS-Based Mobile Navigation System: A Comparison with Maps and Direct Experience." *Journal of Environmental Psychology* 28, no. 1 (March 1, 2008): 74–82. https://doi.org/10.1016/j.jenvp.2007.09.002.

Istomin, Kirill V., and Mark J. Dwyer. "Finding the Way: A Critical Discussion of Anthropological Theories of Human Spatial Orientation with Reference to Reindeer Herders of Northeastern Europe and Western Siberia." *Current Anthropology* 50, no. 1 (February 1, 2009): 29–49. https://doi.org/10.1086/595624.

Jack, Gordon. "Place Matters: The Significance of Place Attachments for Children's Well-Being." *British Journal of Social Work* 40, no. 3 (April 1, 2010): 755–71. https://doi.org/10.1093/bjsw/bcn142.

Jeffery, Hanspeter A., Randolf Menzel Mallot, and Nora S. Newcombe. "Animal Navigation—A Synthesis." *Group* 1 (2010).

Jeffery, Kate J., et al. "Animal Navigation—A Synthesis." In *Animal Thinking: Contemporary Issues in Comparative Cognition*, edited by Julia Fischer and Randolf Menzel. Cambridge, MA: The MIT Press, 2011, 51–76.

Jeffery, Kathryn J. "Remembrance of Futures Past." *Trends in Cognitive Sciences* 8, no. 5 (2004): 197–99.

Jones, Philip G. "Norman B. Tindale Obituary." Australian National University, December 1995. https://www.anu.edu.au/linguistics/nash/aust/nbt/obituary.html.

Kaku, Michio. *The Future of the Mind: The Scientific Quest to Understand, Enhance, and Empower the Mind*. First edition. New York: Doubleday, 2014.

Kalluri, Pratyusha, and Patrick Henry Winston. "Inducing Schizophrenia in an Artificially Intelligent Story-Understanding System," MIT, 2017. http://meta-guide.com/natural-language/nlp/nlu/story-understanding-systems.

Kempermann, G., H. G. Kuhn, and F. H. Gage. "More Hippocampal Neurons in Adult Mice Living in an Enriched Environment." *Year Book of Psychiatry and Applied Mental Health* 1998, no. 9 (January 1, 1998): 399–401.

Kempermann, Gerd, George H. Kuhn, and Fred H. Gage. "More Hippocampal Neurons in Adult Mice Living in an Enriched Environment." *Nature* 386 (April 3, 1997): 493–95. https://doi.org/doi:10.1038/386493a0.

Kennedy, Jennifer J. "Harney, William Edward (Bill) (1895–1962)." In *Australian Dictionary of Biography*. Canberra: National Centre of Biography, Australian National University. http://adb.anu.edu.au/biography/harney-william-edward-bill-10428.

Kerwin, Dale. *Aboriginal Dreaming Paths and Trading Routes: The Colonisation of the Australian Economic Landscape*. East Sussex, UK: Sussex Academic Press, 2010.

Keski-Säntti, Jouko, Ulla Lehtonen, Pauli Sivonen, and Ville Vuolanto. "The Drum as Map: Western Knowledge Systems and Northern Indigenous Map Making." *Imago Mundi* 55 (January 1, 2003): 120–25.

Kirby, Peter Wynn. *Boundless Worlds: An Anthropological Approach to Movement*. New York: Berghahn Books, 2009.

Knierim, James J. "From the GPS to HM: Place Cells, Grid Cells, and Memory." *Hippocampus* (April 1, 2015). https://doi.org/10.1002/hipo.22453.

Knight, Will. "A Robot Uses Specific Simulated Brain Cells to Navigate." *MIT Technol-*

ogy Review, 2015. https://www.technologyreview.com/s/542571/a-robot-finds-its -way-using-artificial-gps-brain-cells/.

Konishi, Kyoko, Venkat Bhat, Harrison Banner, Judes Poirier, Ridha Joober, and Véronique D. Bohbot. "APOE2 Is Associated with Spatial Navigational Strategies and Increased Gray Matter in the Hippocampus." *Frontiers in Human Neuroscience* 10 (July 13, 2016). https://doi.org/10.3389/fnhum.2016.00349.

Kovecses, Zoltan. *Language, Mind, and Culture: A Practical Introduction*. New York: Oxford University Press, 2006.

Kraus, Benjamin J., Robert J. Robinson II, John A. White, Howard Eichenbaum, and Michael E. Hasselmo. "Hippocampal 'Time Cells': Time versus Path Integration." *Neuron* 78, no. 6 (June 19, 2013): 1090–101. http://www.sciencedirect.com/science /article/pii/S0896627313003176.

Kübler-Ross, Elisabeth. *The Wheel of Life*. New York: Simon & Schuster, 2012.

Kuhn, Steven L., David A. Raichlen, and Amy E. Clark. "What Moves Us? How Mobility and Movement Are at the Center of Human Evolution." *Evolutionary Anthropology* 25, no. 3 (May 1, 2016): 86–97. https://doi.org/10.1002/evan.21480.

Kumar-Rao, Arati. "The Memory of Wells." *Peepli* (blog), June 21, 2015. http://peepli.org /stories/the-memory-of-wells/.

Kytta, Marketta. "Children's Independent Mobility in Urban, Small Town, and Rural Environments." In *Growing Up in a Changing Urban Landscape*, edited by Ronald Camstra, 41–52. Assen: Van Gorcum, 1997.

Lavenex, Pierre, and Pamela Banta Lavenex. "Building Hippocampal Circuits to Learn and Remember: Insights into the Development of Human Memory." *Behavioural Brain Research* 254 (October 1, 2013): 8–21. https://doi.org/10.1016/j.bbr.2013.02.007.

Lawton, Carol A. "Gender, Spatial Abilities, and Wayfinding." In *Handbook of Gender Research in Psychology*, edited by Joan C. Chrisler and Donald R. McCreary, 317–41. New York: Springer, 2010. https://doi.org/10.1007/978-1-4419-1465-1_16.

Leadbeater, Charles. "Why There's No Place Like Home—for Anyone, Any More." *Aeon*, 2016. https://aeon.co/essays/why-theres-no-place-like-home-for-anyone-any -more.

Lehn, W. H. "The Novaya Zemlya Effect: An Arctic Mirage." *Journal of the Optical Society of America* 69, no. 5 (May 1, 1979): 776. https://doi.org/10.1364/JOSA.69.000776.

León, Marcia S. Ponce de, et al. "Neanderthal Brain Size at Birth Provides Insights into the Evolution of Human Life History." *Proceedings of the National Academy of Sciences* 105, no. 37 (September 16, 2008): 13764–68. https://doi.org/10.1073/pnas .0803917105.

Leshed, Gilly, Theresa Velden, Oya Rieger, Blazej Kot, and Phoebe Sengers. "In-Car GPS Navigation: Engagement with and Disengagement from the Environment." In *Proceedings of the SIGCHI Conference on Human Factors in Computing Systems*, 1675–84. New York: ACM, 2008. https://doi.org/10.1145/1357054.1357316.

Levinson, Stephen C. "Language and Cognition: The Cognitive Consequences of Spatial Description in Guugu Yimithirr." *Journal of Linguistic Anthropology* 7, no. 1 (June 1, 1997): 98–131. https://doi.org/10.1525/jlin.1997.7.1.98.

———. "Language and Space." *Annual Review of Anthropology* 25 (October 1996): 353–82.

———. *Space in Language and Cognition: Explorations in Cognitive Diversity*. Cambridge: Cambridge University Press, 2003.

Levinson, Stephen C., and David P. Wilkins, eds. *Grammars of Space: Explorations in Cognitive Diversity*. Cambridge: Cambridge University Press, 2006.

Lewis, David. "Memory and Intelligence in Navigation: Review of *East Is a Big Bird*. Gladwin Thomas. Harvard University Press." *Journal of Navigation* 24, no. 3 (July 1971): 423–24. https://doi.org/10.1017/S0373463300048426.

———. "Observations on Route Finding and Spatial Orientation among the Aboriginal Peoples of the Western Desert Region of Central Australia." *Oceania* 46, no. 4 (June 1, 1976): 249–82. https://doi.org/10.1002/j.1834-4461.1976.tb01254.x.

———. "Route Finding and Spatial Orientation." *Oceania* 46, no. 4 (1975): 249–82.

———. "Route Finding by Desert Aborigines in Australia." *Journal of Navigation* 29, no. 1 (January 1976): 21–38. https://doi.org/10.1017/S0373463300043307.

Lewis, David, Curriculum Development Centre, and Aboriginal Arts Board. "The Way of the Nomad." In *From Earlier Fleets: Hemisphere—An Aboriginal Anthology*, edited by Kenneth Russell Henderson, 78–82 Canberra: Australian Government, 1978.

Lewis, G. Malcolm. "Maps, Mapmaking, and Map Use by Native North Americans." In *The History of Cartography: Cartography in the Traditional African, American, Arctic, Australian, and Pacific Societies*, edited by David Woodward and G. Malcolm Lewis, vol. 2, book 3, 51–182. Chicago: University of Chicago Press, 1998.

Liebenberg, Louis. *The Art of Tracking: The Origin of Science*. First edition. Claremont, South Africa: New Africa Books, 2012.

———. *The Origin of Science: On the Evolutionary Roots of Science and Its Implications for Self-Education and Citizen Science*. www.cybertracker.org, 2013.

Lincoln, Margarette, ed. *Science and Exploration in the Pacific: European Voyages to the Southern Oceans in the Eighteenth Century*. Woodbridge, UK: Boydell Press, 1998.

Lindbergh, Anne Morrow. "Airliner to Europe." *Harper's Magazine*, September 1948. https://harpers.org/archive/1948/09/airliner-to-europe/.

———. *Listen! The Wind*. San Diego: Harcourt, Brace, 1938.

Londberg, Max. "KC to STL in 20 Minutes? System That Could Threaten Speed of Sound May Come to Missouri." *Kansas City Star*, April 7, 2017. http://www.kansascity.com/news/local/article143315884.html.

Looser, Diana. "Oceanic Imaginaries and Waterworlds: Vaka Moana on the Sea and Stage." *Theatre Journal* 67, no. 3 (2015): 465–86. https://doi.org/10.1353/tj.2015.0080.

Lord, Albert B. *The Singer of Tales: Third Edition*, edited by David F. Elmer. Center for Hellenic Studies, Washington D.C., 2018.

lostTesla (@lostTesla). "What Is a Sparrow." Tweet. October 1, 2017. https://twitter.com/LostTesla/status/923979654704369664.

Lovis, William A., and Robert Whallon. *Marking the Land: Hunter-Gatherer Creation of Meaning in Their Environment*. New York: Routledge, 2016.

Low, Sam. *Hawaiki Rising: Hōkūle'a, Nainoa Thompson, and the Hawaiian Renaissance*. Waipahu, HI: Island Heritage, 2013.

MacDonald, John. *The Arctic Sky: Inuit Astronomy, Star Lore, and Legend*. First edition. Toronto: Royal Ontario Museum, 1998.

Madsen, Heather Bronwyn, and Jee Hyun Kim. "Ontogeny of Memory: An Update on 40 Years of Work on Infantile Amnesia." In "Developmental Regulation of Memory in Anxiety and Addiction." Special issue. *Behavioural Brain Research* 298, part A (February 1, 2016): 4–14. https://doi.org/10.1016/j.bbr.2015.07.030.

Maguire, Eleanor A., et al. "Navigation-Related Structural Change in the Hippocampi of Taxi Drivers." *Proceedings of the National Academy of Sciences* 97, no. 8 (April 11, 2000): 4398–403. https://doi.org/10.1073/pnas.070039597.

Maguire, Eleanor A., Elizabeth R. Valentine, John M. Wilding, and Narinder Kapur. "Routes to Remembering: The Brains behind Superior Memory." *Nature Neuroscience* 6, no. 1 (January 2003): 90–95. https://doi.org/10.1038/nn988.

Maguire, Eleanor A., Katherine Woollett, and Hugo J. Spiers. "London Taxi Drivers and Bus Drivers: A Structural MRI and Neuropsychological Analysis." *Hippocampus* 16, no. 12 (December 1, 2006): 1091–101. https://doi.org/10.1002/hipo.20233.

Majid, Asifa, Melissa Bowerman, Sotaro Kita, Daniel B. M. Haun, and Stephen C. Levinson. "Can Language Restructure Cognition? The Case for Space." *Trends in Cognitive Sciences* 8, no. 3 (March 2004): 108–14. https://doi.org/10.1016/j.tics.2004.01.003.

Malafouris, Lambros, and Colin Renfrew. *How Things Shape the Mind: A Theory of Material Engagement*. Cambridge, MA: MIT Press, 2013.

Malaurie, Jean. *The Last Kings of Thule: A Year among the Polar Eskimos of Greenland*. Springfield, OH: Crowell, 1956.

———. *Ultima Thulé: Explorers and Natives of the Polar North*. New York: W. W. Norton, 2003.

Markowitsch, Hans J., and Angelica Staniloiu. "Memory, Time and Autonoetic Consciousness." *Procedia—Social and Behavioral Sciences* 126, International Conference on Timing and Time Perception, March 31–April 3, 2014, Corfu, Greece (March 21, 2014): 271–72. https://doi.org/10.1016/j.sbspro.2014.02.406.

Marozzi, Elizabeth, and Kathryn J. Jeffery. "Place, Space and Memory Cells." *Current Biology* 22, no. 22 (2012): R939–42.

Mary-Rousselière, Guy. *Qitdlarssuaq, the Story of a Polar Migration*. Winnipeg: Wuerz, 1991. http://www.worldcat.org/title/qitdlarssuaq-the-story-of-a-polar-migration/oclc/24960667.

Matthews, Luke J., and Paul M. Butler. "Novelty-Seeking DRD4 Polymorphisms Are Associated with Human Migration Distance Out-of-Africa after Controlling for Neutral Population Gene Structure." *American Journal of Physical Anthropology* 3, no. 145 (June 14, 2011): 382–89. https://doi.org/10.1002/ajpa.21507.

Maynard, Micheline. "Prefer to Sit by the Window, Aisle or ATM?" *The Lede*, 2008, 1214616788. https://thelede.blogs.nytimes.com/2008/06/27/prefer-to-sit-by-the-window-aisle-or-atm/.

Mazzullo, Nuccio, and Tim Ingold. "Being Along: Place, Time and Movement among Sámi People." In *Mobility and Place: Enacting Northern European Peripheries*, edited by Jørgen Ole Bærenholdt and Brynhild Granås, 27–38. Farnham: Ashgate, 2012.

McBryde, Isabel. "Travellers in Storied Landscapes: A Case Study in Exchanges and Heritage." *Aboriginal History* 24 (2000): 152–74.

McCann, W. H. "Nostalgia: A Review of the Literature." *Psychological Bulletin* 38, no. 3 (March 1, 1941): 165–82.

McDermott, James. *Martin Frobisher: Elizabethan Privateer*. New Haven, CT: Yale University Press, 2001.

McGhee, Robert. *The Arctic Voyages of Martin Frobisher: An Elizabethan Adventure*. Seattle: McGill-Queen's University Press, 2001.

McLeman, Robert A. *Climate and Human Migration: Past Experiences, Future Challenges*. First edition. New York: Cambridge University Press, 2013.

McLuhan, Marshall. *Understanding Media: The Extensions of Man*. Reprint edition. Cambridge, MA: MIT Press, 1994.

McNaughton, Bruce L., Francesco P. Battaglia, Ole Jensen, Edvard I. Moser, and May-Britt Moser. "Path Integration and the Neural Basis of the 'Cognitive Map.'" *Nature Reviews Neuroscience* 7, no. 8 (August 2006): 663–78. https://doi.org/10.1038/nrn1932.

Menzel, Randolf, et al. "Honey Bees Navigate According to a Map-Like Spatial Memory." *Proceedings of the National Academy of Sciences* 102, no. 8 (February 22, 2005): 3040–45. https://doi.org/10.1073/pnas.0408550102.

Merlan, Francesca. *Caging the Rainbow: Places, Politics and Aborigines in a North Australian Town*. Honolulu: University of Hawaii Press, 1998.

———. *A Grammar of Wardaman*. Berlin: De Gruyter Mouton, 1993.

Milford, Michael John. *Robot Navigation from Nature: Simultaneous Localisation, Mapping, and Path Planning Based on Hippocampal Models*. 2008 edition. Berlin: Springer, 2008.

Milton Freeman Research Limited. *Inuit Land Use and Occupancy Project: A Report*. Ottawa: Minister of Supply and Services, 1976.

Mooney, Chris. "In Greenland's Northernmost Village, a Melting Arctic Threatens the Age-Old Hunt." *Washington Post*, April 29, 2017, sec. Business. https://www.washingtonpost.com/business/economy/in-greenlands-northernmost-village-a-melting-arctic-threatens-the-age-old-hunt/2017/04/29/764ba9be-1bb3-11e7-bcc2-7d1a0973e7b2_story.html.

Moran, Barbara. "The Joy of Driving without GPS." *Boston Globe*, August 8, 2017. https://

www.bostonglobe.com/magazine/2017/08/08/the-joy-driving-without-gps /W36dJaTGw05YFdzyixhj3M/story.html.

Moser, May-Britt, David C. Rowland, and Edvard I. Moser. "Place Cells, Grid Cells, and Memory." *Cold Spring Harbor Perspectives in Biology* 7, no. 2 (February 1, 2015): a021808. https://doi.org/10.1101/cshperspect.a021808.

Mullally, Sinéad L., and Eleanor A. Maguire. "Learning to Remember: The Early Ontogeny of Episodic Memory." *Developmental Cognitive Neuroscience* 9, no. 100 (July 2014): 12. https://doi.org/10.1016/j.dcn.2013.12.006.

———. "Memory, Imagination, and Predicting the Future." *Neuroscientist* 20, no. 3 (June 2014): 220–34. https://doi.org/10.1177/1073858413495091.

Mulvaney, D. J. "Stanner, William Edward (Bill) (1905–1981)." In *Australian Dictionary of Biography*. Canberra: National Centre of Biography, Australian National University, n.d. http://adb.anu.edu.au/biography/stanner-william-edward-bill-15541.

Murray, Elisabeth, Steven Wise, and Kim Graham. *The Evolution of Memory Systems: Ancestors, Anatomy, and Adaptations*. First edition. New York: Oxford University Press, 2017.

Musk, Elon (@elonmusk). "When You Want Your Car to Return, Tap Summon on Your Phone. It Will Eventually Find You Even If You Are on the Other Side of the Country." Tweet. October 3, 2016. https://twitter.com/elonmusk/status/789022017311735808 ?lang=en.

Myers, Fred. "Ontologies of the Image and Economies of Exchange." *American Ethnologist* 31, no. 1 (February 1, 2004): 5–20. https://doi.org/10.1525/ae.2004.31.1.5.

———. *Pintupi Country, Pintupi Self: Sentiment, Place, and Politics among Western Desert Aborigines*. Berkeley: University of California Press, 1991.

Nadel, Lynn. "The Hippocampus and Space Revisited." *Hippocampus* 1, no. 3 (July 3, 1991): 221–29. https://doi.org/10.1002/hipo.450010302.

Nadel, Lynn, and Stuart Zola-Morgan. "Infantile Amnesia: A Neurobiological Perspective." In *Infant Memory: Its Relation to Normal and Pathological Memory in Humans and Other Animals*, vol. 9, edited by Morris Moscovitch, 145. Berlin: Springer, 2012.

Nelson, Richard K. *Hunters of the Northern Ice*. Chicago: University of Chicago Press, 1972.

Newell, Alison, Peter Nuttall, and Elisabeth Holland. "Sustainable Sea Transport for the Pacific Islands: The Obvious Way Forward." In *Future Ship Powering Options: Exploring Alternative Methods of Ship Propulsion*. London: Royal Academy of Engineering, 2013.

Nicholls, Christine Judith. "'Dreamtime' and 'The Dreaming'—An Introduction." *The Conversation*, 2014. http://theconversation.com/dreamtime-and-the-dreaming-an -introduction-20833.

———. "'Dreamtime' and 'The Dreaming': Who Dreamed Up These Terms?" *The Conversation*, January 28, 2014. http://theconversation.com/dreamtime-and-the-dreaming -who-dreamed-up-these-terms-20835.

Niffenegger, Audrey. *Her Fearful Symmetry: A Novel*. New York: Simon & Schuster, 2009.

Norris, Ray P. "Dawes Review 5: Australian Aboriginal Astronomy and Navigation." *Publications of the Astronomical Society of Australia* 33 (2016): e039. https://doi.org/10 .1017/pasa.2016.25.

Norris, Ray P., and Bill Yidumduma Harney. "Songlines and Navigation in Wardaman and Other Australian Aboriginal Cultures." *Journal of Astronomical History and Heritage* 17, no. 2 (April 9, 2014). http://arxiv.org/abs/1404.2361.

Nunn, Patrick D., and Nicholas J. Reid. "Aboriginal Memories of Inundation of the Australian Coast Dating from More Than 7000 Years Ago." *Australian Geographer* 47, no. 1 (September 7, 2015): 11–47. http://www.tandfonline.com/doi/abs/10.1080 /00049182.2015.1077539.

Nuttall, Mark, ed. *Encyclopedia of the Arctic.* First edition. New York: Routledge, 2004.

Nuttall, Peter, Paul D'Arcy, and Colin Philp. "Waqa Tabu—Sacred Ships: The Fijian Drua." *International Journal of Maritime History* 26, no. 3 (August 1, 2014): 427–50. https://doi.org/10.1177/0843871414542736.

Nuttall, Peter Roger. "Sailing for Sustainability: The Potential of Sail Technology as an Adaptation Tool for Oceania; A Voyage of Inquiry and Interrogation through the Lens of a Fijian Case Study." PhD thesis, Victoria University of Wellington, New Zealand, 2013.

O'Grady, Cathleen. "Spatial Reasoning Is Only Partly Explained by General Intelligence." *Ars Technica*, February 24, 2017. https://arstechnica.com/science/2017/02/twin-study-finds-that-spatial-ability-is-more-than-just-intelligence/.

O'Keefe, John. "Biographical." The Nobel Foundation, 2014. https://www.nobelprize.org/nobel_prizes/medicine/laureates/2014/okeefe-bio.html.

O'Keefe, J., and J. Dostrovsky. "The Hippocampus as a Spatial Map: Preliminary Evidence from Unit Activity in the Freely-Moving Rat." *Brain Research* 34, no. 1 (November 1971): 171–75.

O'Keefe, John, and Lynn Nadel. *The Hippocampus as a Cognitive Map.* Oxford/New York: Clarendon Press/Oxford University Press, 1978.

Oudgenoeg-Paz, Ora, Paul P. M. Leseman, and M. (Chiel) J. M. Volman. "Can Infant Self-Locomotion and Spatial Exploration Predict Spatial Memory at School Age?" *European Journal of Developmental Psychology* 11, no. 1 (January 2, 2014): 36–48. https://doi.org/10.1080/17405629.2013.803470.

Palmer, Jason. "Human Eye Protein Senses Earth's Magnetism." BBC News, 2011. http://www.bbc.com/news/science-environment-13809144.

Parush, Avi, Shir Ahuvia, and Ido Erev. "Degradation in Spatial Knowledge Acquisition When Using Automatic Navigation Systems." In *Spatial Information Theory*, 238–54. Lecture Notes in Computer Science. Berlin: Springer, 2007. https://doi.org/10.1007/978-3-540-74788-8_15.

Patzke, Nina, Olatunbosun Olaleye, Mark Haagensen, Patrick R. Hof, Amadi O. Ihunwo, and Paul R. Manger. "Organization and Chemical Neuroanatomy of the African Elephant (Loxodonta Africana) Hippocampus." *Brain Structure & Function* 219, no. 5 (September 2014): 1587–601. https://doi.org/10.1007/s00429-013-0587-6.

Pellegrini, Anthony D., Danielle Dupuis, and Peter K. Smith. "Play in Evolution and Development." *Developmental Review* 27, no. 2 (June 1, 2007): 261–76. https://doi.org/10.1016/j.dr.2006.09.001.

Perkins, Hetti. *Art Plus Soul.* Carlton, Victoria: The Miegunyah Press, 2010.

Peters, Roger. "Cognitive Maps in Wolves and Men." *Environmental Design Research* 2 (1973): 247–53.

Piaget, Jean. *The Construction of Reality in the Child.* New York: Routledge, 2013.

Piaget, Jean, and Barbel Inhelder. *The Child's Conception of Space.* Translated by F. J. Langdon and J. L. Lunzer. New York: W. W. Norton, 1967.

Pilling, Arnold R. "Review of *Review of Aboriginal Tribes of Australia: Their Terrain, Environmental Controls; Distribution, Limits, and Proper Names*, by Norman B. Tindale." *Ethnohistory* 21, no. 2 (1974): 169–71. https://doi.org/10.2307/480950.

Plumert, Jodie M., and John P. Spencer. *The Emerging Spatial Mind.* New York: Oxford University Press, 2007.

Pollack, Lisa. "Historical Series: Magnetic Sense of Birds." www.ks.uiuc.edu, July 1, 2012. http://www.ks.uiuc.edu/History/magnetoreception/#related.

Portugali, Juval. *The Construction of Cognitive Maps.* Berlin: Springer Science & Business Media, 1996.

Praag, Henriette van, Brian R. Christie, Terrence J. Sejnowski, and Fred H. Gage. "Running Enhances Neurogenesis, Learning, and Long-Term Potentiation in Mice." *Proceedings of the National Academy of Sciences* 96, no. 23 (November 9, 1999): 13427–31.

Pravosudov, Vladimir V., and Timothy C. Roth II. "Cognitive Ecology of Food Hoarding: The Evolution of Spatial Memory and the Hippocampus." *Annual Review of Ecology, Evolution, and Systematics* 44, no. 1 (2013): 173–93. https://doi.org/10.1146/annurev-ecolsys-110512-135904.

Prinz, Jesse. "Culture and Cognitive Science." In *The Stanford Encyclopedia of Philosophy*, edited by Edward N. Zalta, Fall 2016. Metaphysics Research Lab, Stanford University, 2016. https://plato.stanford.edu/archives/fall2016/entries/culture-cogsci/.

"Putuparri Tom Lawford: Oral History." Canning Stock Route Project, 2014. http://mira.canningstockrouteproject.com/node/3060. Accessed February 10, 2016.

Qumaq, Taamusi. "A Survival Manual: Annaumajjutiksat." *Tumivut*, no. 78 (1995).

Rasmussen, Knud. *Across Arctic America: Narrative of the Fifth Thule Expedition*. First edition. Fairbanks: University of Alaska Press, 1999.

Redish, A. David. *Beyond the Cognitive Map: From Place Cells to Episodic Memory*. A Bradford Book. Cambridge, MA: MIT Press, 1999.

Relph, Edward. *Place and Placelessness*. London: Pion, 1976.

Revell, Grant, and Jill Milroy. "Aboriginal Story Systems: Re-Mapping the West, Knowing Country, Sharing Space." *Occasion: Interdisciplinary Studies in the Humanities* 5 (March 1, 2013). http://occasion.stanford.edu/node/123.

Richards, Graham. *Race, Racism and Psychology: Towards a Reflexive History*. New York: Routledge, 2003.

Rissotto, Antonella, and Francesco Tonucci. "Freedom of Movement and Environmental Knowledge in Elementary School Children." *Journal of Environmental Psychology* 22, no. 1 (March 1, 2002): 65–77. https://doi.org/10.1006/jevp.2002.0243.

Ritz, Thorsten, Salih Adem, and Klaus Schulten. "A Model for Photoreceptor-Based Magnetoreception in Birds." *Biophysical Journal* 78, no. 2 (February 1, 2000): 707–18. https://doi.org/10.1016/S0006-3495(00)76629-X.

Robaey, Philippe, Sam McKenzie, Russel Schachar, Michel Boivin, and Véronique D. Bohbot. "Stop and Look! Evidence for a Bias Towards Virtual Navigation Response Strategies in Children with ADHD Symptoms." In "Developmental Regulation of Memory in Anxiety and Addiction." Special issue. *Behavioural Brain Research* 298, part A (February 1, 2016): 48–54. https://doi.org/10.1016/j.bbr.2015.08.019.

Robbins, J. "GPS Navigation . . . but What Is It Doing to Us?" In *2010 IEEE International Symposium on Technology and Society*, 309–18. Wollongong, Australia: IEEE, 2010. https://doi.org/10.1109/ISTAS.2010.5514623.

———. "When Smart Is Not: Technology and Michio Kaku's the Future of the Mind [Leading Edge]." *IEEE Technology and Society Magazine* 35, no. 2 (June 2016): 29–31. https://doi.org/10.1109/MTS.2016.2554439.

Roberts, Mere. "Mind Maps of the Maori." *GeoJournal* 77, no. 6 (September 4, 2010): 741–51. https://doi.org/10.1007/s10708-010-9383-5.

Rogers, Dallas. "The Poetics of Cartography and Habitation: Home as a Repository of Memories." *Housing, Theory and Society* 30, no. 3 (September 1, 2013): 262–80. https://doi.org/10.1080/14036096.2013.797019.

Rose, Deborah Bird. *Dingo Makes Us Human: Life and Land in an Australian Aboriginal Culture*. First edition. Cambridge: Cambridge University Press, 2000.

Rosello, Oscar (Rosello Gil). "NeverMind: An Interface for Human Memory Augmentation." Thesis, Massachusetts Institute of Technology, 2017. http://dspace.mit.edu/handle/1721.1/111494.

Rozhok, Andrii. *Orientation and Navigation in Vertebrates*. 2008 edition. Berlin: Springer, 2010.

Rubin, David C. *Memory in Oral Traditions: The Cognitive Psychology of Epic, Ballads, and Counting-Out Rhymes*. New York: Oxford University Press, 1997.

———. "The Basic-Systems Model of Episodic Memory." *Perspectives on Psychological Science* 1, no. 4 (December 2006): 277–311.

Rubin, David C., and Sharda Umanath. "Event Memory: A Theory of Memory for Laboratory, Autobiographical, and Fictional Events." *Psychological Review* 122, no. 1 (2015): 1–23.

Rundstrom, Robert A. "A Cultural Interpretation of Inuit Map Accuracy." *Geographical Review* 80, no. 2 (1990): 155–68. https://doi.org/10.2307/215479.

Sacks, Oliver. *The Island of the Colorblind*. First edition. New York: Vintage, 1998.

Saint-Exupéry, Antoine de. *Wind, Sand and Stars*. Translated by Lewis Galantiere. San Diego: Harcourt, 2002.

"Satnavs 'Switch Off' Parts of the Brain." University College London News, March 21, 2017. http://www.ucl.ac.uk/news/news-articles/0317/210317-satnav-brain-hippocampus.

Schacter, Daniel L., Donna Rose Addis, Demis Hassabis, Victoria C. Martin, R. Nathan Spreng, and Karl Szpunar. "The Future of Memory: Remembering, Imagining, and the Brain." *Neuron* 76, no. 4 (2012): 677–94. https://dash.harvard.edu/handle/1/11688796.

Schilder, Brian M., Brenda J. Bradley, and Chet C. Sherwood. "The Evolution of the Human Hippocampus and Neuroplasticity." 86th Annual Meeting of the American Association of Physical Anthropologists, 2017. http://meeting.physanth.org/program/2017/session25/schilder-2017-the-evolution-of-the-human-hippocampus-and-neuroplasticity.html.

Schiller, Daniela, et al. "Memory and Space: Towards an Understanding of the Cognitive Map." *Journal of Neuroscience* 35, no. 41 (October 14, 2015): 13904–11. https://doi.org/10.1523/JNEUROSCI.2618-15.2015.

Scott, James C. *Against the Grain: A Deep History of the Earliest States*. New Haven, CT: Yale University Press, 2017.

Scott, Laurence. *The Four-Dimensional Human: Ways of Being in the Digital World*. New York: W. W. Norton, 2016.

Sebti, Bassam. "4 Smartphone Tools Syrian Refugees Use to Arrive in Europe Safely." *Voices*, February 17, 2016. https://blogs.worldbank.org/voices/4-smartphone-tools-Syrian-refugees-use-to-arrive-in-Europe-safely.

Seymour, Julie, Abigail Hackett, and Lisa Procter. *Children's Spatialities: Embodiment, Emotion and Agency*. Berlin: Springer, 2016.

Sharp, Andrew. *Ancient Voyagers in the Pacific*. Paper edition. New York: Penguin, 1957.

Shaw-Williams, Kim. "The Social Trackways Theory of the Evolution of Language." *Biological Theory* 12, no. 4 (December 1, 2017): 195–210. https://doi.org/10.1007/s13752-017-0278-2.

———. "The Triggering Track-Ways Theory." Master's thesis, Victoria University of Wellington, 2011. http://researcharchive.vuw.ac.nz/handle/10063/1967.

Short, John Rennie. *Globalization, Modernity and the City*. New York: Routledge, 2013.

Siewers, Alfred K. "Colors of the Winds, Landscapes of Creation." In *Strange Beauty: Ecocritical Approaches to Early Medieval Landscape*, 97–110. The New Middle Ages. New York: Palgrave Macmillan, 2009. https://doi.org/10.1057/9780230100527_4.

Simmons, Matilda. "The Ocean Keepers." *Fiji Times Online*, December 5, 2016. http://www.fijitimes.com/story.aspx?id=380883.

Singal, Jesse. "How the Brains of 'Memory Athletes' Are Different." *Science of Us*, 2017. http://nymag.com/scienceofus/2017/03/how-the-brains-of-memory-athletes-are-different.html.

Smith, Catherine Delano. "The Emergence of 'Maps' in European Rock Art: A Prehistoric Preoccupation with Place." *Imago Mundi* 34, no. 1 (January 1, 1982): 9–25. https://doi.org/10.1080/03085698208592537.

Snead, James E., Clark L. Erickson, and J. Andrew Darling, eds. *Landscapes of Movement: Trails, Paths, and Roads in Anthropological Perspective*. Philadelphia: University of Pennsylvania Press, 2009. http://www.jstor.org/stable/j.ctt3fhjb3.

Souman, Jan L., Ilja Frissen, Manish N. Sreenivasa, and Marc O. Ernst. "Walking Straight into Circles." *Current Biology* 19, no. 18 (September 2009): 1538–42. https://doi.org /10.1016/j.cub.2009.07.053.

Spiers, Hugo J., and Caswell Barry. "Neural Systems Supporting Navigation." *Current Opinion in Behavioral Sciences* 1, no. 1 (2015): 47–55. https://doi.org/10.1016/j.cobeha .2014.08.005.

Spiers, Hugo J., and Eleanor A. Maguire. "Thoughts, Behaviour, and Brain Dynamics during Navigation in the Real World." *NeuroImage* 31, no. 4 (July 15, 2006): 1826–40. https://doi.org/10.1016/j.neuroimage.2006.01.037.

Spiers, Hugo J., Eleanor A. Maguire, and Neil Burgess. "Hippocampal Amnesia." *Neurocase* 7, no. 5 (January 1, 2001): 357–82. https://doi.org/10.1076/neur.7.5.357.16245.

Squire, Larry R. "The Legacy of Patient H.M. for Neuroscience." *Neuron* 61, no. 1 (January 15, 2009): 6–9. https://doi.org/10.1016/j.neuron.2008.12.023.

Squire, Larry R., Anna S. van der Horst, Susan G. R. McDuff, Jennifer C. Frascino, Ramona O. Hopkins, and Kristin N. Mauldin. "Role of the Hippocampus in Remembering the Past and Imagining the Future." *Proceedings of the National Academy of Sciences* 107, no. 44 (November 2, 2010): 19044–48. https://doi.org/10.1073/pnas .1014391107.

Stea, David, James M. Blaut, and Jennifer Stephens. "Mapping as a Cultural Universal." In *The Construction of Cognitive Maps*, 345–60. GeoJournal Library. Dordrecht: Springer, 1996. https://doi.org/10.1007/978-0-585-33485-1_15.

Stern, Pamela R., and Lisa Stevenson. *Critical Inuit Studies: An Anthology of Contemporary Arctic Ethnography*. Lincoln: University of Nebraska Press, 2006.

Stilgoe, John R. *Landscape and Images*. Charlottesville: University of Virginia Press, 2015.

———. *Outside Lies Magic: Regaining History and Awareness in Everyday Places*. London: Bloomsbury USA, 2009.

———. *What Is Landscape?* Cambridge, MA: MIT Press, 2015.

"Story: Aboriginal Guides." Canning Stock Route Project, 2012. http://www.canningstock routeproject.com/history/story-aboriginal-guides/.

Suddendorf, Thomas, Donna Rose Addis, and Michael C. Corballis. "Mental Time Travel and the Shaping of the Human Mind." *Philosophical Transactions of the Royal Society B: Biological Sciences* 364, no. 1521 (May 12, 2009): 1317–24. https://doi.org/10 .1098/rstb.2008.0301.

Sugiyama, Lawrence S., and Michelle Scalise Sugiyama. "Humanized Topography: Storytelling as a Wayfinding Strategy." *Amerian Anthropologist*, n.d.

Sugiyama, Michelle Scalise. "Food, Foragers, and Folklore: The Role of Narrative in Human Subsistence." *Evolution and Human Behavior* 22, no. 4 (July 1, 2001): 221–40. https://doi.org/10.1016/S1090-5138(01)00063-0.

———. "Oral Storytelling as Evidence of Pedagogy in Forager Societies." *Frontiers in Psychology* 8 (March 29, 2017). https://doi.org/10.3389/fpsyg.2017.00471.

Sutton, Peter. "Aboriginal Maps and Plans." In *The History of Cartography: Cartography in the Traditional African, American, Arctic, Australian, and Pacific Societies*, edited by David Woodward and G. Malcolm Lewis, vol. 2, 387–416. Chicago: University of Chicago Press, 1998.

———, ed. *Dreamings: The Art of Aboriginal Australia*. First edition. New York: George Braziller, 1997.

Tandy, C. A. "Children's Diminishing Play Space: A Study of Inter-Generational Change in Children's Use of Their Neighbourhoods." *Australian Geographical Studies* 37, no. 2 (July 1, 1999): 154–64. https://doi.org/10.1111/1467-8470.00076.

Teit, James Alexander. *Traditions of the Thompson River Indians of British Columbia*. American Folk-Lore Society. Boston: Houghton, Mifflin, 1898.

Teki, Sundeep, et al. "Navigating the Auditory Scene: An Expert Role for the Hippocampus." *Journal of Neuroscience* 32, no. 35 (August 29, 2012): 12251–57. https://doi.org /10.1523/JNEUROSCI.0082-12.2012.

"The Magnetic Sense Is More Complex Than Iron Bits." *Evolution News*, April 2016. https://evolutionnews.org/2016/04/the_magnetic_se/.

Thomson, Helen. "Cells That Help You Find Your Way Identified in Humans." *New Scientist*, August 4, 2013. https://www.newscientist.com/article/dn23986-cells-that-help-you-find-your-way-identified-in-humans/.

Tolman, Edward C. "Cognitive Maps in Rats and Men." *Psychological Review* 55, no. 4 (July 1948): 189–208. https://doi.org/10.1037/h0061626.

Traoré, Genome Res, et al. "Genetic Clues to Dispersal in Human Populations: Retracing the Past from the Present." *Genetics* 145 (1997): 505.

Tuan, Yi-Fu. *Topophilia: A Study of Environmental Perception, Attitudes, and Values*. Reprint edition. New York: Columbia University Press, 1990.

Tulving, Endel. "Episodic Memory and Common Sense: How Far Apart?" *Philosophical Transactions of the Royal Society of London. Series B, Biological Sciences* 356, no. 1413 (September 29, 2001): 1505–15. https://doi.org/10.1098/rstb.2001.0937.

Turnbull, David, and Helen Watson. *Maps Are Territories: Science Is an Atlas; A Portfolio of Exhibits*. Chicago: University of Chicago Press, 1989.

Vanhoenacker, Mark. *Skyfaring: A Journey with a Pilot*. Reprint edition. New York: Vintage, 2016.

Vargha-Khadem, F., D. G. Gadian, K. E. Watkins, A. Connelly, W. Van Paesschen, and M. Mishkin. "Differential Effects of Early Hippocampal Pathology on Episodic and Semantic Memory." *Science* 277, no. 5324 (July 18, 1997): 376–80. https://doi.org/10.1126/science.277.5324.376.

Viard, Armelle, et al. "Mental Time Travel into the Past and the Future in Healthy Aged Adults: An fMRI Study." *Brain and Cognition* 75, no. 1 (February 1, 2011): 1–9. https://doi.org/10.1016/j.bandc.2010.10.009.

Vito, Stefania de, and Sergio Della Sala. "Predicting the Future." *Cortex* 47, no. 8 (September 1, 2011): 1018–22. https://doi.org/10.1016/j.cortex.2011.02.020.

Vleck, Jenifer van. *Empire of the Air*. Cambridge, MA: Harvard University Press, 2013.

von Frisch, Karl. *Bees: Their Vision, Chemical Senses, and Language*. Ithaca, NY: Cornell University Press, 2014.

Vycinas, Vincent. *Earth and Gods: An Introduction to the Philosophy of Martin Heidegger*. The Hague: Martinus Nijhoff, 1961.

Wachowich, Nancy, Apphia Agalakti Awa, Rhoda Kaukjak Katsak, and Sandra Pikujak Katsak. *Saqiyuq: Stories from the Lives of Three Inuit Women*. Montreal: McGill-Queen's University Press, 2001.

Walker, M. "Navigating Oceans and Cultures: Polynesian and European Navigation Systems in the Late Eighteenth Century." *Journal of the Royal Society of New Zealand* 42, no. 2 (May 28, 2012): 93–98. http://www.tandfonline.com/doi/abs/10.1080/03036758.2012.673494#.V0ip25MrI6g.

"Walking in Circles: Scientists from Tubingen Show That People Really Walk in Circles When Lost." Max-Planck-Gesellschaft, August 20, 2009. https://www.mpg.de/596269/pressRelease200908171.

Wang, Ranxiao Frances, and Elizabeth S. Spelke. "Human Spatial Representation: Insights from Animals." *Trends in Cognitive Sciences* 6, no. 9 (September 1, 2002): 376–82. https://doi.org/10.1016/S1364-6613(02)01961-7.

Wegman, Joost, Anna Tyborowska, Martine Hoogman, Alejandro Arias Vásquez, and Gabriele Janzen. "The Brain-Derived Neurotrophic Factor Val66Met Polymorphism Affects Encoding of Object Locations during Active Navigation." *European Journal of Neuroscience* 45, no. 12 (June 2017): 1501–11. https://doi.org/10.1111/ejn.13416.

Wehner, Rüdiger. "Desert Ant Navigation: How Miniature Brains Solve Complex Tasks." *Journal of Comparative Physiology A* 189, no. 8 (July 23, 2003): 579–88. https://doi.org/10.1007/s00359-003-0431-1.

———. "On the Brink of Introducing Sensory Ecology: Felix Santschi (1872–1940)—

Tabib-En-Neml." *Behavioral Ecology and Sociobiology* 27, no. 4 (October 1, 1990): 295–306. https://doi.org/10.1007/BF00164903.

Wehner, R., and S. Wehner. "Insect Navigation: Use of Maps or Ariadne's Thread?" *Ethology Ecology & Evolution* 2, no. 1 (May 1, 1990): 27–48. https://doi.org/10.1080 /08927014.1990.9525492.

Weil, Simone, with an introduction by T. S. Eliot. *The Need for Roots: Prelude to a Declaration of Duties towards Mankind.* First edition. London: Routledge, 2001.

Weltfish, Gene. *The Lost Universe: Pawnee Life and Culture.* Lincoln: University of Nebraska Press, 1977.

White, April. "The Intrepid '20s Women Who Formed an All-Female Global Exploration Society." *Atlas Obscura*, April 12, 2017. http://www.atlasobscura.com/articles /society-of-woman-geographers.

Widlok, Thomas. "Landscape Unbounded: Space, Place, and Orientation in ≠Akhoe Hai// Om and Beyond." *Language Sciences* 2–3, no. 30 (2008): 362–80. https://doi.org/10 .1016/j.langsci.2006.12.002.

———. "Orientation in the Wild: The Shared Cognition of Hai‖om Bushpeople." *Journal of the Royal Anthropological Institute* 3, no. 2 (1997): 317–32. https://doi.org/10.2307 /3035022.

———. "The Social Relationships of Changing Hai‖om Hunter Gatherers in Northern Namibia, 1990–1994." London: London School of Economics and Political Science, 1994.

Will, Udo. "Oral Memory in Australian Aboriginal Song Performance and the Parry-Kirk Debate: A Cognitive Ethnomusicological Perspective." *Proceedings of the International Study Group on Music Archaeology* 10 (2000): 1–29.

Winston, Patrick Henry. "The Strong Story Hypothesis and the Directed Perception Hypothesis." In *AAAI Fall Symposium: Advances in Cognitive Systems*, 2011.

Winston, Patrick Henry, and Dylan Holmes. *The Genesis Manifesto: Story Understanding and Human Intelligence.* Draft, 2017.

Wolbers, Thomas, and Mary Hegarty. "What Determines Our Navigational Abilities?" *Trends in Cognitive Sciences* 14, no. 3 (February 6, 2010). http://www.sciencedirect .com/science/article/pii/S1364661310000021.

Woodward, David, and G. Malcolm Lewis, eds. *The History of Cartography: Cartography in the Traditional African, American, Arctic, Australian, and Pacific Societies.* First edition. Chicago: University of Chicago Press, 1998.

Woolley, Helen E., and Elizabeth Griffin. "Decreasing Experiences of Home Range, Outdoor Spaces, Activities and Companions: Changes across Three Generations in Sheffield in North England." *Children's Geographies* 13, no. 6 (November 2, 2015): 677–91. https://doi.org/10.1080/14733285.2014.952186.

Wositzky, Jan. *Born under the Paperbark Tree: A Man's Life.* Edited by Yidumduma Bill Harney. Revised edition. Marleston, South Australia: JB Books, 1998.

Wrangel, Ferdinand Petrovich Baron. *Narrative of an Expedition to the Polar Sea, in the Years 1820, 1821, 1822 & 1823. Commanded by Lieutenant, Now Admiral Ferdinand Von Wrangel.* Edited by Edward Sabine. New York: Harper and Bros., 1841.

Yarlott, Wolfgang, and Victor Hayden. "Old Man Coyote Stories: Cross-Cultural Story Understanding in the Genesis Story Understanding System." Thesis, Massachusetts Institute of Technology, 2014. http://dspace.mit.edu/handle/1721.1/91880.

Yates, Frances. *The Art of Memory.* London: Random House, 2014.

"Yiwarra Kuju." National Museum of Australia. http://www.nma.gov.au/education/resources /units_of_work/yiwarra_kuju.

Zalucki, Myron P., and Jan H. Lammers. "Dispersal and Egg Shortfall in Monarch Butterflies: What Happens When the Matrix Is Cleaned Up?" *Ecological Entomology* 35, no. 1 (February 1, 2010): 84–91. https://doi.org/10.1111/j.1365-2311.2009.01160.x.

Zucker, Halle R., and Charan Ranganath. "Navigating the Human Hippocampus without a GPS." *Hippocampus* 25, no. 6 (June 1, 2015): 697–703. https://doi.org/10.1002 /hipo.22447.

INDEX